The Institute of Biology's
Studies in Biology no 56

The Biology of
Slime Moulds

J. M. Ashworth

Ph.D., Professor of Biology, University of Essex

and

Jennifer Dee

Ph.D. Senior Lecturer, Genetics Department, University of Leicester

Edward Arnold

First published 1975
by Edward Arnold (Publishers) Limited
25 Hill Street, London W1X 8LL

Boards edition ISBN: 0 7131 2511 X
Paper edition   ISBN: 0 7131 2512 8

Printed in Great Britain by
Butler & Tanner Ltd, Frome and London

General Preface to the Series

It is no longer possible for one textbook to cover the whole field of Biology and to remain sufficiently up-to-date. At the same time teachers and students at school, college or university need to keep abreast of recent trends and know where significant developments are taking place.

To meet the need for this progressive approach the Institute of Biology has for some years sponsored this series of booklets dealing with subjects specially selected by a panel of editors. The enthusiastic acceptance of the series by teachers and students at school, college and university shows the usefulness of the books in providing a clear and up-to-date coverage of topics, particularly in areas of research and changing views.

Among features of the series are the attention given to methods, the inclusion of a selected list of books for further reading and, wherever possible, suggestions for practical work.

Readers' comments will be welcomed by the authors or the Education Officer of the Institute.

1975

The Institute of Biology,
41 Queen's Gate
London, SW7 5HU

Preface

Slime moulds are not only intriguing organisms in themselves; they are extremely useful in current research. In laboratories all over the world they are being used to investigate fundamental biological problems such as cell differentiation, cell movement and the controls of cell growth and nuclear division. This book is written by two authors because there are two rather different groups of slime moulds which are being used to tackle these different basic problems. Chapter 1 was written jointly; Chapters 2–4 on the acellular slime moulds by Jennifer Dee and Chapters 5–7 on the cellular slime moulds by John Ashworth. What we have written is not a comprehensive text but is very much a personal view of the slime moulds arising from our own research experience and interests. We hope that we may stimulate readers to take an active interest in these fascinating organisms and perhaps to follow investigations of their own.

We should like to thank all those who have helped us with the book, in particular C. A. Beasley, Jim Mackley, Ian Riddell and P. Paters for their assistance with the illustrations. We are indebted to all our colleagues and research associates for many stimulating and enjoyable discussions.

Essex and
Leicester 1975

J.M.A.
J.D.

Contents

1 General Introduction

We are often asked, by our students, the question 'Why do you study slime moulds?' and you are no doubt thinking 'Why should I read a book about such obscure creatures?' There are really two kinds of answer that can be given to questions of this sort. The first is that all forms of life are interesting and since we find slime moulds not only interesting but beautiful we study them. This, however, is an answer which only really committed biologists seem to find adequate and so we usually go on, as we will in this book, to explain that an increasing number of biologists are studying the slime moulds because they appear to show, in an apparently uncomplicated way, many of the phenomena associated with the cell differentiation processes which characterize the so-called 'higher' animals and plants.

Living organisms can be divided, quite clearly, into two basic types. The cells of the prokaryotes (bacteria and blue-green algae) have a relatively simple structure with no organized chromosomal arrangement of their DNA, no nuclear membrane and thus no clearly defined nucleus. The cells of the eukaryotes (which comprise all living things other than bacteria, viruses and blue-green algae) have, by contrast, a rather complex structure with many discrete sub-cellular organelles including a clearly defined nucleus with DNA organized in chromosomes. The lower or so-called 'primitive' eukaryotes consist only of a single cell and thus only one cell type but the bodies of 'higher' eukaryotes consist of assemblies of many different cells and many different cell types. Despite this complexity, all higher eukaryotes, at some stage of their life cycle, consist of a single cell and thus at some point in the formation of the adult body the biochemical and structural characteristics of the progeny of this single cell must diverge, leading to the differentiation of one cell type from another.

The slime moulds, or Mycetozoa, are eukaryotes which show this basic life cycle in what is, perhaps, its simplest form. They all have a uninucleate amoeboid stage in their life cycle from which can develop a multinucleate or multicellular stage which can, in turn, give rise to a fruiting body containing dormant spores (i.e. the amoebae differentiate). The fruiting bodies resemble those produced by the lower fungi and this combination of protozoan and fungal characteristics has led to great taxonomic confusion since de Bary first suggested the name Mycetozoa (mycet = fungal; zoa = animal) for these organisms in 1859. Most classifications nowadays include the Mycetozoa amongst the lower fungi but the true evolutionary relationship of the slime moulds to other organisms remains uncertain. However, this combination of a genuinely microbial amoeboid stage in their life cycle and a clear-cut differentiation process is just what is attracting biologists to these organisms at the moment.

The life cycle of the acellular slime moulds (Myxomycetes) comprises a unicellular amoeboid stage, a multinucleate 'plasmodium' stage and a sporulation stage (Figs. 1–1, 2–1). The processes by which the amoebae form plasmodia differ in different species and are in many cases unknown. In all cases, however, the plasmodium is a syncytium with mil-

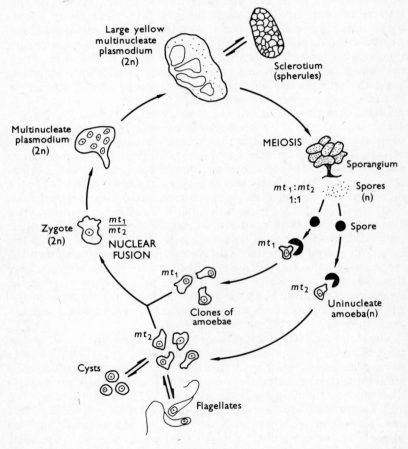

Fig. 1–1 The life-cycle of a Myxomycete, *Physarum polycephalum*. This is a heterothallic form in which plasmodia are formed by the mating of amoebae of different 'mating-types'. (mt_1, mt_2) (n = haploid, 2n = diploid).

lions of nuclei sharing a common cytoplasm which can be several centimetres or even a metre or so in diameter with no permanent static form. The stimuli which lead the plasmodium to form fruiting bodies also differ in different species, but lack of food and changes in light intensity are often important. The fruiting bodies are dry, brittle and immobile structures which contain thousands of uninucleate spores, each of which is surrounded by a thick cell wall which renders it resistant to drought. The spores can be readily dispersed by wind and rain. When moistened each spore germinates to give rise to a uninucleate amoeba which feeds on bacteria, grows, multiplies and thus completes the life cycle.

The cellular slime moulds (Acrasiales) resemble the Myxomycetes in having a microscopic, amoeboid phase in their life cycle but these amoebae *never* fuse to form a multinucleate plasmodium. Instead when the amoebae run out of (or are deprived of) their bacterial food they converge (Fig. 5–1) about a central point to form a multicellular aggregate. By analogy with the Myxomycetes this has been called a 'pseudoplasmodium' but the term 'grex' is better since the cellular slime mould aggregate is nothing like a plasmodium. The grex is motile, with a definite shape and after a period of time (Chapter 5) stops moving, rounds up, and develops into a fruiting body. The component cells of the grex become either the stalk or the spore cells of the fruiting body (Fig. 5–1). The spores, like those of the Myxomycetes, are surrounded by a thick cell wall, which renders them resistant to desiccation, and are dispersed by wind and rain. The life cycle is completed when each spore germinates, in the presence of bacteria, to give rise to another amoeba.

Thus the two major groups of slime moulds have superficially similar life cycles but differ in that one, the Myxomycetes, can form a multinucleate plasmodium stage, whereas the other, the Acrasiales, is characterized by the fact that at all stages the individual cells retain their separate identity.

Both Myxomycetes and Acrasiales originally came to the notice of biologists because they are so common in nature. Their habitat is the soil and rotting vegetation such as leaves and logs. The fruiting bodies appear above the surface of the substrate and can sometimes be seen and collected from such material. Also, when samples of soil, humus or bark are taken into the laboratory and placed on nutrient media, slime mould amoebae or plasmodia frequently emerge.

Two species in particular have been used as experimental organisms, the Myxomycete *Physarum polycephalum* and the cellular slime mould *Dictyostelium discoideum*. They have been used for different reasons, as we shall show, and both have been subjected to a variety of different experimental attacks, by the techniques of biochemistry, genetics and cytology. We hope to demonstrate to some extent how these different approaches are used and how they can be advantageously combined to study particular processes in a single orgamism.

Since we, the authors, are both deeply involved in experimental work with *P. polycephalum* and *D. discoideum* respectively, our approach to the slime moulds as a group is inevitably biased. We hope, however, that it is not unbalanced. Although our major interest is in using slime moulds to study processes of general biological significance, we have not forgotten that they are also organisms. They not only survive in natural conditions outside our laboratories but have apparently done so for a very long time with great success. Thus the features which they possess are presumably those that are valuable to them in natural conditions. It is often important to remember this when interpreting observations on a particular organism and attempting to generalize from them. Conversely, one must not forget that during long laboratory culture, features may become established by selection that are quite different from those occurring under the pressures of natural selection in the organism's usual habitat.

At a time of growing enthusiasm for slime moulds as 'tools' in biological research, which we admittedly share, we hope that this book will not only explain the reasons for this enthusiasm but also help to maintain a balanced approach to these interesting organisms, by considering their biology as a whole. We will also suggest some areas where you might yourself experiment on slime moulds to fill in some of the many gaps in our knowledge about them.

2 Detailed Studies on Myxomycete Plasmodia

2.1 The plasmodium

Let us suppose we have isolated for culture a plasmodium of the Myxomycete, *Physarum polycephalum*. It is growing on an agar-based complete medium in a Petri dish and has the appearance shown in Fig. 2–1a. It is bright yellow and appears moist, and when touched it easily squashes into a yellow formless smear. It has a characteristic smell which many people find quite pleasant. If we examine it by transmitted light under a low-power microscope, we can see that it consists of a network of moist branching channels. Focussing on the contents of the channels, we can detect a rapid streaming movement in which thousands of particles are jostled along in the stream in an apparently disorganized way. If we watch for long enough, however, we will begin to detect some regularity. After about half a minute, the stream slows down, almost stops, reverses its direction, and then with rapidly gathering speed, flows back in the direction from which it came. These reversals of flow are repeated at regular intervals, although if we follow the course of adjacent channels, we will see that reversals do not occur in all of them at the same time. The plasmodium itself does not remain stationary, and if we look at its edge, we can usually detect its gradual advance across the agar medium. The advancing edge consists of a continuous sheet of protoplasm, laced with small veins in which there is a slow trickling of fluid and particles. Behind the advancing edge, the network is composed of bigger, well-defined veins with open spaces between them. Since the general form of the plasmodium is maintained as it grows and moves across the plate, the pattern of the veins in any one area is constantly changing. We may see veins join one another and enlarge as their flow increases or shrink and disappear as their contents flow elsewhere.

2.2 Protoplasmic streaming

Protoplasmic streaming is a characteristic of many cells but the streaming we see in Myxomycete plasmodia is on an unusually massive scale. The plasmodium has therefore been regarded as favourable material in which to study the mechanism of streaming. Many different hypotheses have been put forward, but there is now much evidence to

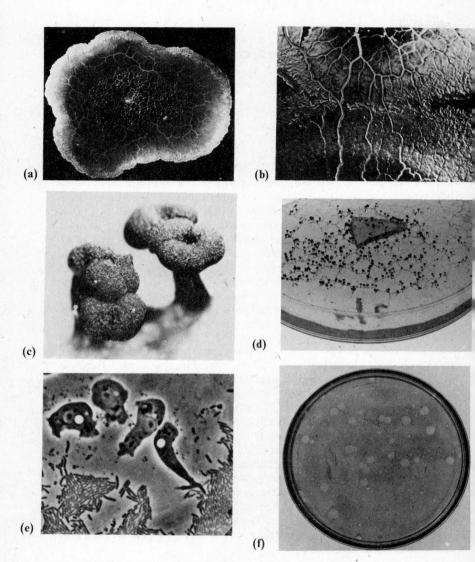

Fig. 2–1 *Physarum polycephalum.* **(a)** A growing plasmodium on agar medium (× 0.7) (From DEE, J. and POULTER, R. T. M. (1970). *Genet. Res. Camb.*, **15**, 35–41.) **(b)** Fusion of two plasmodia (× 4). (From CARLILE, M. J. and DEE, J. (1967). *Nature, Lond.*, **215**, 832–4.) **(c)** Sporangia (× 20). *P. polycephalum* usually forms multiple lobes on each stalk, as its name suggests, but sometimes it forms single-headed sporangia as in **(d)**. **(d)** Sporangia on a plate after plasmodial growth on agar medium (× 0.8). Remains of plasmodial veins are visible but they contain no living protoplasm. **(e)** Amoebae feeding on *Escherichia coli* (× 1000). Contractile vacuoles appear white and circular. The single nucleus (light with dark central nucleolus) can be seen clearly in three amoebae. **(f)** Amoebal 'plaques' on a lawn of bacteria, (× 0.5). (From DEE, J. (1962). *Genet. Res. Camb.*, **3**, 11–23.)

support the idea that the motive force of streaming in a plasmodium is provided by contractile proteins resembling those of vertebrate muscle. Both actin and myosin have been extracted from *P. polycephalum* plasmodia and, when purified, these proteins form complexes with muscle myosin and muscle actin respectively. The complexes have similar contractile properties to muscle actomyosin. Actomyosin extracted directly from the plasmodium or formed by mixing purified *Physarum* actin and myosin is also contractile. Contraction in these systems is promoted by the addition of ATP, which is thought also to be the source of energy for streaming, since the local application of ATP to a plasmodium stimulates the rate of streaming. It is still uncertain whether the ATP is generated only aerobically in a plasmodium or whether some anaerobic formation can also take place. Most observations suggest that a plasmodium can continue streaming for only a short time in the absence of oxygen.

Studies of plasmodial structure with the electron microscope have frequently revealed the presence of fibrils in or near the clear cytoplasm which forms the walls of the veins and there is some suggestive evidence that these fibrils may be composed of actin and myosin, but this has not yet been conclusively demonstrated. It is of course an attractive idea that waves of contraction in such fibres squeeze the protoplasm along the veins, but most observers have been unable to detect any such movements of the walls. Thus we cannot conclude yet that the mechanism of protoplasmic streaming in the plasmodium is understood. There also remains the very interesting question of how streaming is coordinated so that a plasmodium of any size maintains a roughly constant form during growth and migration and does not normally fragment into pieces.

2.3   Plasmodial migration

If we cut out a small piece (say 1 cm^2) of a plasmodium growing on an agar medium and lift it, together with the underlying agar, onto a fresh plate of nutrient medium, we will find after a few hours that the plasmodium is growing off the agar block onto the fresh medium, and spreading out evenly as it grows. If, however, we make such a transfer to a non-nutrient medium, such as plain water agar, we will find, after 12–24 hours, that the plasmodium is not spreading evenly around the block but is moving as a discrete organism across the surface of the agar. The plasmodium is now moving as a result of protoplasmic streaming in the absence of growth, its movement is coordinated so that it migrates as a whole around the plate, and it continues to do this for several days until it eventually dies, finds a food source on which it can grow, or forms a resistant sclerotium (Fig. 1–1). As it moves around, it leaves a 'slimetrack', consisting of polygalactose slime which is secreted by the plasmodium, probably as a protection against drying, and this allows us to trace the path of its movements (Fig. 4–1d).

Although it would clearly be interesting to investigate whether various external factors influence the direction of plasmodial migration, few such experiments have been reported. One set of tests by M. J. Carlile was designed to detect chemotaxis towards carbohydrate solutions. A piece of plasmodium of *P. polycephalum* was placed at the centre of an agar plate a few centimetres equidistant from two wells, one containing the solution to be tested and the other containing water. The test was repeated several times with each solution and if the plasmodium moved directly towards the carbohydrate well in a significant number of tests, it was concluded that positive chemotaxis was occurring. Such chemotaxis was demonstrated towards some sugars, such as glucose, maltose and galactose, but not towards others, such as fructose and sucrose. It was found also that the former sugars supported plasmodial growth and the latter did not. Many more experiments of this type could be done, testing not only nutrients and other chemicals, but such factors as light, humidity and temperature. It would also be interesting to study the reaction of plasmodia to other organisms such as bacteria, yeasts and fungi, many of which they have been observed to feed upon, and also their reactions to inert particles, such as carbon, which they will freely ingest.

2.4   Plasmodial fusion

When two pieces of the same plasmodium touch, they will normally fuse. This may be observed by cutting two blocks from a plasmodium and placing them on nutrient medium a few centimetres apart. After about 24 hours, the plasmodia will have formed two fans spreading across the plate. When they are a few millimetres apart, they should be continuously observed through the microscope since fusion takes place very rapidly after the plasmodial membranes touch. The first discernible sign of fusion is usually a slow trickling of particles from one plasmodium to the other, and soon after this, synchronous reversals of flow in adjacent veins of the two plasmodia can be detected. A rapidly increasing flow of fluid and particles is then seen in the region where the membranes first met, this being channelled into a few large veins which seem to materialize and grow before our eyes (Fig. 2–1b). Within a few hours of contact, the two plasmodia may appear as one.

The ability of two plasmodia to fuse on contact has been found to be under genetic control. It has often been noted that plasmodia of different species or of different isolates of the same species fail to fuse. An analysis of the situation has been carried out in two species, *P. polyce-*

phalum and *Didymium iridis,* and the results in both are similar. In each of these species, several genetic loci control plasmodial fusion and only plasmodia which carry the same alleles at all these loci fuse with each other. Isolates of the same species from different geographical regions usually carry different alleles and therefore do not fuse, and even within an isolate there seems to be much heterogeneity, so that if progeny plasmodia are bred from a single strain in the laboratory, many of them will not fuse. The genetic analysis involved in these studies is complex and beyond the scope of this book. When two non-identical plasmodia do fuse there is often a rapid lethal interaction, which may result in the death of one or both of the plasmodia. Comparisons have been drawn between this process and incompatibility reactions in other organisms, but it is not yet clear that these are justified. However, since there are rather few systems in which incompatibility can be studied, it certainly seems worthwhile that some investigation of these processes in plasmodia should be attempted.

The ecological significance of genes controlling plasmodial fusion is unknown. It must be emphasized that fusion between genetically different plasmodia is not an essential part of the sexual cycle of Myxomycetes. Sexual fusion or 'mating' takes place between genetically different *amoebae,* and a separate locus ('mating type') is known to control this process in both *P. polycephalum* and *D. iridis* (Chapter 4). The 'plasmodial fusion loci' do not influence this process and since they reduce the chance of fusion between dissimilar plasmodia, they will tend to reduce gene exchange. Whether this is important to the organism or not is unknown.

In the laboratory, however, the ready fusion of genetically similar plasmodia has been useful in many experiments. One can fuse pieces of the same plasmodium which have been given different treatments or which are in different states and observe the behaviour of the fusion product. This type of experiment is suggested to the reader as providing another means of investigating the effects of external factors on plasmodia.

2.5   Plasmodial constituents

Attempts to study microscopically the morphology of a plasmodium such as that of *P. polycephalum* are usually not very informative. There appears to be an outer layer of slime which may be fibrillar in structure. Within this there is a membrane bounding the protoplasm. The outer protoplasm ('ectoplasm') is clearer and more viscous than the inner protoplasm ('endoplasm'). It is the granular endoplasm which flows in the veins. Fibrils which are perhaps involved in promoting protoplasmic streaming lie in or near the ectoplasm. Probably the ectoplasm and endoplasm ('plasmagel' and 'plasmasol') can change into one another quite readily, allowing the rapid growth and diminution of veins which is observed in some conditions: in this way the plasmodium resembles a giant amoeba. The contents of the endoplasm in the veins are nuclei, mitochondria, ribosomes, pigment granules and many 'particles', 'inclusions' and 'vacuoles' whose nature and function are unknown. If the plasmodium is fed on particulate food, such as oat flakes or bacteria, many 'food vacuoles' will be seen containing ingested particles. Plasmodia fed on soluble or liquid media also contain many vacuoles however.

Although cytology has not been very illuminating, much has been learnt by the extraction and biochemical study of plasmodial constituents. Since the plasmodium is virtually one large cell, it is possible to homogenize it and to isolate samples of nuclei, nucleoli, mitochondria and other subcellular fractions. Studies of such isolated fractions have contributed much to the work that is described below.

2.6   Synchronous mitosis

Although mitosis was first seen in plasmodial nuclei around 1900, the first extensive study was done in 1932 by Howard. He examined samples of a growing *P. polycephalum* plasmodium at 15 minute intervals over a period of hours, and found that mitosis occurred synchronously in all the nuclei. This large scale natural synchrony, which involves many millions of nuclei, has offered an exciting system in which to study many aspects of mitosis and the 'cell cycle'. About 1953, Dr H. P. Rusch in the McArdle laboratory for Cancer Research in Wisconsin initiated a programme of research to study mitosis and DNA synthesis in *P. polycephalum* and this research has continued and developed since then in many laboratories throughout the world.

2.7   The cell cycle

The cell cycle is the period between successive divisions of a cell. In eukaryotes, this period is said to consist of four stages: mitosis (M), which is usually accompanied by cell division, DNA synthesis (S) and the intervals G_1 preceding DNA synthesis and G_2 following DNA synthesis. (See Fig. 2–2.) When attention is focussed on the cytological events of the cell cycle, we tend to place great emphasis on the behaviour of the chromosomes during mitosis and to regard the rest of the cycle as an undifferentiated 'interphase', during which the chromosomes are diffuse and nothing can be seen to happen. Biochemical studies have shown, however, that a great deal is happening during interphase and that some of the events during this time are as precisely sequenced as

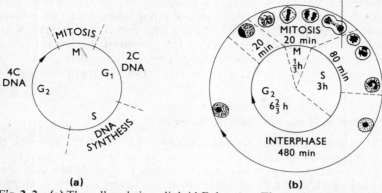

Fig. 2–2 **(a)** The cell cycle in a diploid Eukaryote. The nuclear DNA content doubles in S and halves in M. **(b)** The cell cycle in *P. polycephalum*. Inner ring shows duration of M, S. and G_2. Outer ring shows changes in appearance of nuclei (see text). Dense spot in nucleus during interphase is the nucleolus which moves to the side of the nucleus and then disappears during mitosis. (Redrawn from data and figures in MOHBERG, J. and RUSCH, H. P. (1969), *J. Bacteriol.* **97**, 1411.)

those visible during mitosis. If one is to understand the controls of cell and nuclear division, it is essential to study these events. Although it is possible to study single cells, in some organisms these do not yield sufficient material for biochemical analyses, and it is desirable to have populations of cells in synchrony so that a large quantity of material at each stage of the cell cycle may be obtained. Methods of synchronizing cells are known, but there is always a risk that these may disturb the processes we wish to study. *P. polycephalum* plasmodia are valuable because they show natural mitotic synchrony. Of course, a Myxomycete plasmodium is not ideal material in which to study the regulation of cell division in animal cells, which is the main concern in understanding cancer. Although there is nuclear division, there is no cell division, and the organism apparently has no mechanisms to regulate its overall size. In a differentiated, multicellular organism, many controls must have evolved to regulate the growth of the organism as a whole and to balance the multiplication of the various cell types, and these are presumably based on interactions between cells. Thus there will be many questions we cannot answer by studying a Myxomycete plasmodium. However the unique advantage of the plasmodium is in fact due to its syncytial structure. Synchronous nuclear division is a general characteristic of syncytia and is a result of the nuclei being in a common cytoplasm. Although the plasmodium is an unusually large syncytium its cytoplasm is also unusually well-mixed, as a result of the vigorous protoplasmic streaming. This implies of course that the timing of mitosis depends on factors in the cytoplasm and is not controlled by an autonomous nuclear 'clock'. One might in any case expect this to be so, since nuclear divisions must be related to the growth of the cell if balance is to be maintained. Thus by studying a Myxomycete plasmodium, we may at least learn something about the control of nuclear division.

2.8   Studies of the cell cycle in *Physarum*

Particular techniques have been developed to study the cell cycle in *P. polycephalum* plasmodia. The plasmodia are grown in quantity in liquid shaken cultures on a synthetic medium. Fully-defined media may be used, in which the chemical composition of every nutrient is known, or richer 'complete media' in which, for example, peptone is provided instead of amino acids. The culture is mechanically shaken to provide constant aeration, since the plasmodium is aerobic and will not grow when deeply submerged. Growth is logarithmic, with a doubling time of about 10 hours in rich medium. In liquid, shaken culture, plasmodia grow in the form of 'microplasmodia', each less than a millimetre in diameter and containing perhaps a few hundred nuclei. The nuclei in a microplasmodium are in synchrony but those in different microplasmodia are not. To obtain a fully synchronized culture, the microplasmodia are harvested and pippetted onto a moist filter paper laid on a grid over liquid medium. As they spread on this surface, they meet and fuse, and soon a single surface plasmodium is formed and a few hours later the nuclei become synchronous. Most experiments are carried out at the third or fourth cycle after fusion, so that there is time for all cellular processes to become fully synchronized. In a Petri dish culture one may use a plasmodium about 7 cm in diameter, which contains an estimated number of 10^8 nuclei in good synchrony. The difference in timing of events between different nuclei is only a few minutes and the total time between successive mitoses is about 10 hours.

The approach of mitosis is indicated by visible changes in the nuclei (Fig. 2–2). The nucleolus, which is large and well-defined during interphase, breaks down and disappears a few minutes before mitosis is seen. The usual stages of mitosis can be recognized, the whole process taking about 20 minutes from prophase to telophase. Since the nuclear membrane remains visible throughout, this is termed 'intranuclear' or 'closed' mitosis and contrasts with the 'open' mitosis seen in Myxomycete amoebae and in most other eukaryote cells. For about an hour and a half after telophase, the nucleolus undergoes visible reconstruction. In a culture with a 10 hour doubling time, the next mitosis will occur some 8 hours after the end of reconstruction. Thus processes can be timed in relation to mitosis.

Synthesis of most of the DNA in the plasmodium was found to begin immediately after telophase and to continue for about 3 hours and this DNA can be 'labelled' by supplying radioactive or high-density precursors of DNA during the S period. There is no G_1 period in growing plasmodia.

If label is supplied during G_2, it is found that some DNA becomes labelled and is therefore being synthesized at this time. As a result of a number of experiments, this has been identified as mitochondrial DNA and ribosomal DNA. Mitochondrial DNA is of course situated in the cytoplasm and can be studied in isolated mitochondria. Ribosomal DNA is situated inside the nuclei, mostly in the nucleoli, and consists of the genes which code for ribosomal RNA. This has been studied in a number of ways. Ribosomal DNA is of higher density than the rest of the nuclear DNA, and therefore forms a 'satellite' peak in density gradient centrifugation; it binds with ribosomal RNA because of its complementary base sequence; it consists of many replicates of the same base sequences and it can be extracted from isolated nucleoli. These properties agree with those known for ribosomal DNA in other organisms, and *Physarum* offers a favourable system in which to study them further.

The synthesis of RNA through the cell cycle has also been the object of many studies and a repeatable pattern is found. There is little synthesis during mitosis and two peaks of synthesis are found in interphase, the first during S and the second in G_2. Many attempts have been made to identify and distinguish the types of RNA made at these times, and there is some evidence that the first peak may consist largely of messenger RNA synthesis and the second of ribosomal RNA. Both DNA synthesis and RNA synthesis may also be studied in isolated nuclei, which show the same pattern of synthesis as the plasmodium from which they are taken.

The activity of some particular enzymes has been followed and is found to vary repeatably through the cycle. Some experiments have been designed to test the dependence of mitosis and DNA synthesis on protein synthesis occurring at particular times in the cycle. The results suggest that proteins necessary for DNA synthesis are synthesized during S and that proteins necessary for mitosis are synthesized up to a few minutes before metaphase.

This very brief account of cell cycle studies in *Physarum* indicates the extensive knowledge that is being acquired about the complex processes involved. However an understanding of the ultimate controls of the cell cycle, which must be operating to give balanced growth, probably requires a rather different approach.

2.9   The control of mitosis

Attempts to elucidate the control of the cell cycle are based on the idea that DNA synthesis and nuclear division must be linked to cell growth so that a constant nuclear : cytoplasmic ratio is maintained. Many experiments demonstrate that the cytoplasm influences the timing of mitosis. When microplasmodia coalesce, their nuclei soon become synchronous. If two plasmodia are fused which are at different stages of G_2, their nuclei divide synchronously at a time intermediate between the times at which they would have divided if the plasmodia had remained separated. This experiment suggests a linear change during G_2 of some factor in the cytoplasm which stimulates or inhibits mitosis.

Many complex theoretical models for control of mitosis and DNA synthesis have been proposed, and efforts are being made to distinguish between these experimentally. Most experiments attempt to disturb the normal cycle of events by altering the nuclear : cytoplasmic ratio, for example by destroying some of the nuclei by irradiation. Observations are also made at different growth rates by using different media. It is hoped that soon it may become possible to isolate strains carrying gene mutations which affect the control of the cell cycle. If the cycle is controlled ultimately by the genes, it should be possible to induce mutations in these genes, and each one might be expected to interfere with the cycle in a precise way, thus revealing the function of the normal gene. Techniques for isolating and analysing mutants are now available (see Chapter 4) but the development of genetics in *P. polycephalum* has unfortunately lagged far behind the biochemical studies.

3 Variation and Differentiation of Myxomycete Plasmodia

3.1 Plasmodial morphology among Myxomycetes

The previous chapter described many of the detailed studies that are being carried out on plasmodia of *P. polycephalum*. No other Myxomycete species has been studied so intensively, and so it is impossible to say whether most of the features described are common to other Myxomycetes. However, many observations have been made in other species, and these are particularly due to the work of Professor C. J. Alexopoulos, and his students in the University of Texas, who have collected and cultured a wide range of Myxomycetes. Myxomycetes are usually classified into six main orders, three of which contain rather few known species. In all, perhaps 500 species are known, and 50–60 of these have been cultured through their whole life cycle in the laboratory. Classification of orders, families, genera and species is based almost entirely on the morphology of the fruiting bodies, which show clearly-defined differences between the groups and remain reasonably stable within each species. However, Professor Alexopoulos has recently pointed out that the form of the plasmodium also differs between the major groups. *Physarum polycephalum* belongs to the order Physarales, and its large, thick, fan-shaped, easily-cultured plasmodium (the 'phaneroplasmodium') is characteristic of this group. Most of the species used in the laboratory for experimental work belong to the Physarales, including the genera *Didymium, Physarella, Badhamia* and *Fuligo*, as well as several species of *Physarum*. The plasmodium is yellow, brown or white in different species. Members of the Physarales are also the Myxomycetes most frequently encountered in nature, being larger and more obvious than the rest.

3.2 Sporulation

All Myxomycetes develop some type of fruiting body, bearing spores, and although these are given various names they appear, with one exception, to be variations on a basic form, the sporangium (Fig. 2–1, c, d). The processes involved in development of the sporangium may generally be termed 'sporulation', and they have been studied cytologically and biochemically. The first obvious morphological change is the development of papillae which rise above the substrate on which the

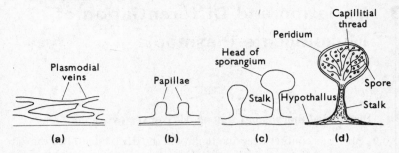

Fig. 3–1 Stages in the formation of sporangia (**a–c**) and diagrammatic section of mature sporangium (**d**). The peridium often ruptures and disappears as the sporangium matures, leaving the spores exposed.

plasmodium has been growing (Fig. 3–1). These appear to form by protoplasmic movement, in the absence of growth, and movement is perpendicular to the substrate, not oriented by gravity. A papilla elongates by movement of protoplasm up a stalk to form a rounded or finger-like head at its tip. The centre of the stalk becomes filled with granules which give it strength, and the stalk is eventually composed entirely of non-living material. Its outer wall is a thick layer continuous with the 'peridium' which is the outer layer of the head, within which the spores will be confined. Inside the head, two developmental processes take place at about the same time but are apparently not closely linked. These are the development of the capillitium, which is an internal framework of non-living threads; and the cleavage of the protoplasm to form spores. The result is a head containing spores which are mingled with capillitial threads and supported by them but are not attached to them.

Formation of the capillitium has been followed in several species and some electron-microscope studies have been done. In some species, it appears that a tubular network is first formed, opening on the surface of the peridium, through which calcium carbonate is moved and deposited inside the tubes and on the peridium. The presence of $CaCO_3$ on the peridium and within the capillitium is a characteristic of many species, particularly in the Physarales. The chemical composition of the capillitial threads is not known, though both chitin and cellulose have been reported to be present.

Cleavage of the protoplasm to form the spores has also been observed cytologically and it is agreed that the result of repeated cleavages is the formation of uninucleate spores. However, there has been much disagreement about whether one or two nuclear divisions accompany cleavage and whether these are meiotic or mitotic divisions. For many years it was generally believed that meiosis took place before spore delineation, so that each spore contained a single haploid product of meiosis. However, recent electron-microscopic studies provide strong evidence that only one mitotic division precedes spore formation and that meiosis occurs within the spore. This might be expected to result in a spore carrying all four products of meiosis, but there is evidence that, at least in some species, three of the four products degenerate. The survival of only one haploid nucleus after meiosis is known to occur in many other organisms, particularly during the formation of animal eggs, so it would not be an entirely unexpected process.

After spore delineation, the spore walls thicken and often change in colour. The markings and colour of the walls differ in different species. In *P. polycephalum*, a rapid colour change occurs from the yellow of the plasmodium to the dark brown or black of the spores, and melanin synthesis is thought to be involved.

Variation among species may be found in the colour and shape of sporangia and in the form of the peridium and capillitium. Some species have no capillitium. Often the sporangia are quite beautiful, consisting of tiny, regularly-shaped heads borne on long slender stalks, and delicately coloured pink, yellow or red (Fig. 3–2). The fruiting bodies of

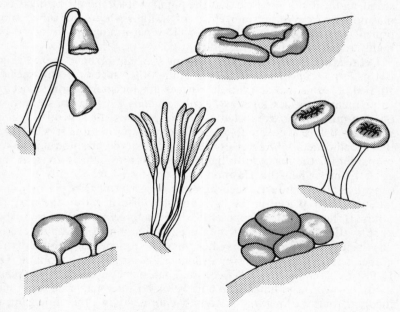

Fig. 3–2   Some of the many different forms found among Myxomycete fruiting bodies. (Re-drawn from LISTER, A. (1925). *A Monograph of the Mycetozoa*, Oxford University Press.)

many species do not have the stalked sporangial form described but appear like sessile (stalkless) sporangia, sometimes closely grouped together to form a large ill-defined mass. These types of fruiting body are given other names, such as 'plasmodiocarp' and 'aethelium', but in general their development and structure, so far as they are understood, resemble those of sporangia.

3.2.1 *Sporulation in controlled conditions*

In a series of experiments first published in 1938, W. D. Gray showed that light was necessary for sporulation of some Myxomycete species. Investigating four species with yellow plasmodia and ten species which were not pigmented, all in conditions of starvation, he found that light was required only by the yellow species and in later studies no exceptions to this rule have been demonstrated. The yellow pigment is therefore regarded as a photoreceptor which, when exposed to light, alters the metabolic activities of the plasmodium to cause the changes observed in sporulation. However, the nature of the pigment, which seems to be very complex, and its role in sporulation are not yet understood, and it is still possible that the pigment itself merely protects the plasmodium from the harmful effects of the light, while some other component acts as the photoreceptor. Yellow plasmodia are found to be highly resistant to U-V irradiation.

Detailed studies of the factors affecting sporulation have been carried out in *P. polycephalum* and also in some other species. In all species studied, it seems that starvation is necessary for sporulation to occur. In pigmented species, such as *P. polycephalum*, a period of starvation on non-nutrient or exhausted medium, perhaps for several days, is necessary before exposure to light will induce sporulation. It is thought that specific metabolic changes occur in the starved plasmodium which prepare it for the reaction to light. If exposure to light is given at the critical time, when the plasmodium is ready to react, only a short exposure (about 1 hour) is necessary to induce fruiting. The plasmodium may then be replaced in the dark and it will fruit within the next 12 hours. If, however, it is exposed to light before the necessary period of starvation has elapsed, it may need to remain in the light several days before fruiting is observed. A precise sequence of treatments to induce sporulation in *P. polycephalum* has been worked out in Dr H. P. Rusch's laboratory in Wisconsin. These can be applied only to plasmodia grown in liquid shaken culture or on filter papers over liquid medium, but the method is valuable for studying the biochemical changes involved, since sporulation can be induced to occur at a precisely predictable time.

3.2.2   *Sporulation as a developmental process*

In several features, sporulation resembles developmental processes in higher organisms. Morphogenesis occurs in the absence of growth and there is synthesis of enzymes and new substances, such as galactosamine and melanin in the spore walls. In addition, a 'point of commitment' is reached when the development becomes irreversibly determined. This occurs 2–4 hours after exposure to light. Before this point, replacing the plasmodium on nutrient medium will result in growth and sporulation will not take place, even though the plasmodium has been exposed to the usual periods of starvation and illumination. After the point of commitment, nutrient medium no longer induces growth, and the plasmodium continues to sporulate. This resembles the 'determination' observed in some embryonic animal cells which continue in their developmental pathway after being removed from the 'inducer'.

If the mechanisms of these developmental processes are similar, the Myxomycete plasmodium may be a convenient system in which to study them, for the same reasons as it is useful for cell cycle studies. A large quantity of protoplasm develops in the same way at the same time, so that biochemical studies of the changes are possible, and when appropriate treatments are used, the developmental changes occur at predictable times.

So far, however, rather little understanding of the processes involved has been achieved. One of the most interesting questions is whether new gene activity is induced at the time of sporulation, since the control of gene activity, at the level of transcription (RNA synthesis) seems likely to be one of the fundamental processes involved in cell differentiation in higher organisms. Experiments utilizing DNA–RNA hybridization suggest that new RNA species are synthesized during sporulation, but the evidence is not yet conclusive. Since gene activity is likely to be controlled by regulatory genes, it will be necessary to use genetic analysis of mutants to understand the systems involved. At present no progress has been made in identifying gene mutations affecting sporulation.

3.2.3   *Sporulation in natural conditions*

Sporulation is an essential stage of the sexual cycle of a Myxomycete, involving the formation of gametes in which genetic reassortment can occur. As in other organisms, the sexual cycle has presumably evolved because it gives variability to an organism faced with a variable environment. In Myxomycetes, spore formation also allows an escape from conditions of starvation, cold or drought. Spores can survive in such unfavourable circumstances for months or even years, and can also be dispersed to new situations by water or wind. Since Myxomycetes occupy different habitats in many different regions of the world, we may expect that the conditions inducing sporulation will vary among different species. Very little is known, however, about sporulation in natural conditions or the factors that influence it. Some species seem to sporulate at particular times of year, and different temperature optima have been reported for different species. Sporulation usually occurs at night in most Myxomycetes, including those with a light requirement. Fruiting bodies are usually found on exposed surfaces, unlike the plasmodia which tend to remain hidden under bark, leaves or soil. Presumably this allows easier dispersal of the spores.

Since even starving plasmodia are capable of movement over considerable distances, one might expect that they would be capable of actively selecting the most favourable situation in which to sporulate. In particular, one might expect to find that pigmented plasmodia would move towards the light at this time. Although this seems an obvious field for investigation, almost no studies have been reported and this is therefore suggested as an area in which the reader might like to experiment. The response of plasmodia to light, gravity, food sources, moisture, and texture of the surface could be investigated, during conditions of growth and during starvation. In *P. polycephalum*, light has been reported to inhibit plasmodial growth on nutrient media and several of the very early (19th century) observers of Myxomycetes reported that they were negatively phototropic. A change to positive phototropism might be expected in starved plasmodia. Two more recent studies have suggested that plasmodia choose a position in which to sporulate. In one, plasmodia of *Badhamia utricularis* (a pigmented species) were exposed to light in flowerpots after feeding on rolled oats in the dark for a week. At first they were observed to move to shady positions in the pots; but later they moved towards the light and fruited on the borderline between light and shade. A gradient of humidity was present since the pots were standing in water, and it is suggested that the plasmodia may also have been responding to this. In the other study (by the author), plasmodia were fed on oats at the bottom of conical flasks in the dark for a week. A lump of cotton wool was then placed on the layer of oats and the flask incubated in the dark for a further 24 hours. After this time, it was almost always observed that the plasmodium had migrated into and through the cotton wool to emerge in a compact mass on its upper side. The flask was then exposed to light and fruiting bodies were formed without further migration of the plasmodium. Since this was merely devised as a method for inducing fruiting, there was no investigation of whether the plasmodium was responding to gravity, a humidity gradient, the fibrous texture of the wool or to other factors, but the movement did not seem to be random.

3.3 Sclerotium formation

If a plasmodium is deprived of food in the absence of light or is slowly dried, it may form a sclerotium, a hard-walled, resistant form which will remain dormant until moistened. The sclerotium forms a brittle or leathery crust on the medium, sometimes reticulate, sometimes a small compact mass. Internally, it is irregularly divided into cells or 'spherules', each containing several nuclei. Spherules may also be induced in liquid cultures of *P. polycephalum* by placing plasmodia in medium without an energy source. Since enzyme synthesis is involved in spherulation as well as morphological change, it has been regarded as an example of differentiation and has been intensively studied in some laboratories. As with sporulation, however, evidence for the synthesis of new classes of RNA during spherulation is not conclusive, and it has not yet been possible to use genetic methods to study the process.

4 Myxomycete Amoebae and Plasmodium Formation

Let us suppose we have a culture of *Physarum polycephalum* which has sporulated. The fruiting bodies appear dry and brittle but may be powdery when touched. If we place some clumps of spores in water and crush them with a glass rod or spatula, we will gradually produce a reddish-coloured suspension. Examination with a microscope shows that many of the spores have been separated although large clumps are still present. The spores are rounded, often variable in size, and little detail can be seen in them. If we allow the suspension to stand for some hours, however, many spores will germinate to release cells which swim actively by means of flagella. Placed on an agar surface, these cells assume an amoeboid form and glide slowly around, sometimes with the flagella still extended in front of them. These myxamoebae, or 'amoebae' as they are often called, are less than 10 μm in diameter and are difficult to observe in detail except with high magnification under a phase contrast microscope. Under these conditions, we can see that each cell has a single nucleus with a large nucleolus and one or more contractile vacuoles (Fig. 2–1e). If provided with bacteria as food, the amoebae ingest them by means of pseudopodia and many food vacuoles containing bacteria in various stages of digestion can be seen in each cell. The cells divide at intervals and if plenty of food is provided, a thriving population can be easily obtained (Fig. 4–1a). When the food supply runs out, the cells encyst and remain in this dormant form until supplied with food.

If we consider the current state of knowledge about *P. polycephalum* myxamoebae, we find a very different situation from that which we have seen for the plasmodium. In the first place, no fully-defined synthetic medium is known on which the amoebae of *P. polycephalum* can be cultured. The amoebae fail to grow on any of the various synthetic media which have been developed for *P. polycephalum* plasmodia. Slow growth of amoebae has recently been achieved in a very complex soluble axenic medium and it is hoped that a defined medium will gradually be developed from this, but for nearly all experiments so far, living or dead bacteria have been used as food. Secondly, almost no biochemical analysis of myxamoebae has been attempted.

These two deficiencies are certainly related to each other. The interest which biochemists felt in analysing *Physarum* plasmodia resulted in intensive efforts to develop a synthetic and defined medium. The success in developing media led in turn to the proliferation of biochemical studies. Since amoebae did not have the attractive feature of synchronous mitosis, few attempts were made to culture them. It is becoming clear, however, that it would be extremely interesting if comparative biochemical studies could be made on the two stages.

Throughout the Myxomycetes, spore germination and the appearance and behaviour of the amoebae are very similar. Flagella formation usually occurs quite rapidly whenever amoebae are placed in water and the flagellated cells are often called 'swarm cells'. A number of physiological and cytological studies have been done on flagella formation but the mechanism of this remarkably rapid transformation is not yet understood.

4.1 Plasmodium formation

When amoebae from a spore sample have been cultured for several days on bacteria on an agar surface, plasmodia begin to appear. They are often first detected under a low-power microscope by the clear areas which they cause in the amoebal lawn. These clear areas are not due to aggregation of the amoebae but to the predatory habits of the plasmodia, which at this stage appear as small rounded blobs within them (Fig. 4–1b, c). Under a higher magnification, we can see that the plasmodia are multinucleate and that they are eating large numbers of amoebae. Slow streaming of the protoplasm can be seen, by which the nuclei and the food vacuoles containing encysted amoebae are churned about. The small rounded plasmodia grow rapidly and assume club-shaped, fan-shaped and eventually reticulate forms (Fig. 4–1, c, d). They fuse readily with other small plasmodia and also undergo nuclear divisions. One or two days after they were first visible, they have the structure of typical plasmodia and may be treated as such.

The above description could probably be applied to most Myxomycetes without modification. However, when we begin to enquire into the earliest stages of development, and ask how multinucleate plasmodia first arise from uninucleate amoebae, we are faced with a rather confusing array of data.

4.2 Heterothallic strains

In several Myxomycete species, cultures of amoebae grown from single spores do not give rise to plasmodia, but when the cultures are mixed together, plasmodia are formed in some of the mixtures (Fig. 4–2). This situation has been found in about five species and has been extensively analysed in *P. polycephalum* and *Didymium iridis*.

The results are explained by assuming that plasmodia are formed

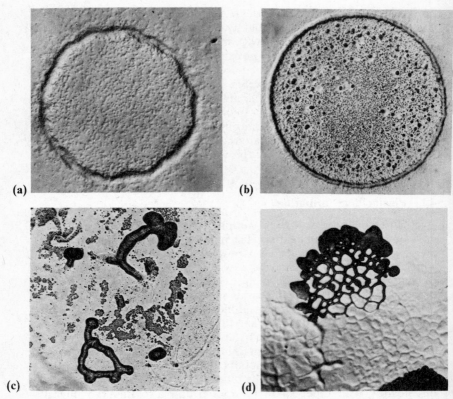

Fig. 4–1   *Physarum polycephalum.* (a) A single colony of heterothallic amoebae on a bacterial lawn (× 25). Amoebae heap up around the periphery as the colony grows. (b) Plasmodia forming in a colony of 'Colonia' amoebae (× 15). Clear light areas are seen where plasmodia (dark blobs) have ingested encysted amoebae. Plasmodia forming in a 'cross' of heterothallic amoebae would appear similar. (c) Young plasmodia (× 20). Most of the encysted amoebae have now been ingested and plasmodial tracks are seen in the clear areas. (From DEE, J. (1966) *J. Protozool.*, **13**, 610–16.) (d) A young plasmodium transferred to nutrient medium migrates away from the inoculum block and leaves slime-tracks on the agar (× 10).

only when amoebae of different 'mating-type' are brought together and that a single spore gives rise to amoebae of only one mating-type. This type of life cycle is called 'heterothallic', a term used in Mycology for fungi which have mycelia of different sexes or mating-types. However, no differences in morphology or physiology have yet been detected in amoebae of different mating-type.

Among the spores from a single plasmodium, only two mating-types appear, and these segregate in equal ratio (1 : 1). When a plasmodium is formed by mixing amoebae of two mating-types, the same two mating-types segregate 1 : 1 among the spores derived from it (Fig. 1–1).

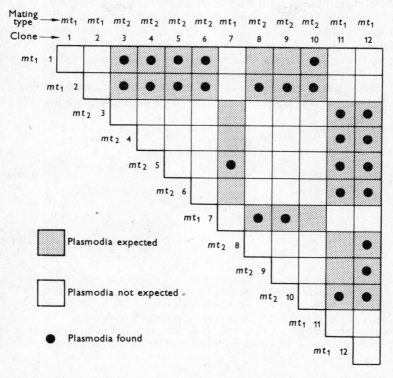

Fig. 4–2   Plasmodium formation found when 12 clones of *P. polycephalum* amoebae are mixed in all pairwise combinations. The pattern found is consistent with that expected if amoebae are of two 'mating-types' (*mt*₁ and *mt*₂). A few tests have failed to give plasmodia where expected but there are no plasmodia where they are *not* expected. (Modified from DEE, J. (1966) *J. Protozool.* **13,** 610.)

These results agree with the idea that mating-types are due to alleles of a chromosomal gene, that amoebae are haploid and therefore carry only one allele each, and that the alleles segregate during meiosis in the spores. If this is true, we must also assume that only one of the four meiotic products usually survives in each spore, although the cytological evidence for this is not conclusive (Chapter 3). In about 2% of single-spore cultures, plasmodia are found, and these may be attributed to the occasional survival of more than one meiotic product in some spores. In the next generation of spores from these plasmodia, the mating-types segregate normally again, as we would expect. Since it appears from these observations that amoebal cultures from single spores are not always pure clones, most studies are now done on cultures from single isolated amoebae. Unlike spores, amoebae cannot easily be isolated by micromanipulation, but they can be 'plated out' to give single colonies which can be assumed to be clones (see Appendix and Fig. 2–1 f).

When plasmodia of the same species are isolated from new geographical regions, they are usually found to carry mating-types different from

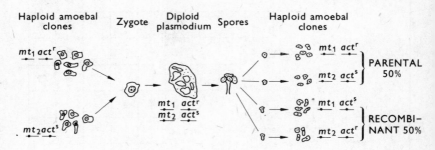

Fig. 4–3 Diagrammatic representation of a cross between two clones of *P. polycephalum* amoebae: one mating-type 1, actidione resistant (mt_1 act^r) and the other mating-type 2, actidione sensitive (mt_2 act^s). Symbols indicate supposed genotypes. Since parental and recombinant types are equally frequent among progeny clones, it is assumed that two unlinked allelic pairs are segregating as shown. (Data from DEE, J. (1966), *Genet. Res., Camb.*, **8**, 101.)

those already isolated. In *P. polycephalum* six mating-types are known, each one being compatible (i.e. forming plasmodia) with the other five. In *D. iridis*, even more types are known. In each species, genetic studies have shown that mating-type is determined by alleles of a single gene (a 'multiple allelic series'). In *P. polycephalum*, the gene has been named *mt* and the alleles mt_1, mt_2, etc.

By using two characters as genetic markers, for example, in *P. polycephalum*, mating-type and resistance to actidione (cycloheximide), it was shown that genetic recombination occurred during the life cycle (Fig. 4–3). 'Recombinant' and 'parental' types were equally frequent among the amoebae from the cross, showing that meiosis had occurred only in diploid, heterozygous nuclei. Thus nuclear fusion must have occurred at some time before spore formation. Since several observers had reported seeing cell and nuclear fusion between pairs of amoebae, it seemed probable that fusion occurred at the beginning of plasmodium formation. The plasmodia were therefore expected to be diploid. Measurements of the nuclear DNA content of amoebae and plasmodia of *P. polycephalum* have confirmed this prediction. By isolation of known numbers of nuclei from several stages of the life cycle and chemical estimation of their total DNA, it has been found that growing (G_2) plasmodia contain twice as much DNA/nucleus as growing (G_2) amoebae.

In heterothallic isolates of *P. polycephalum* and *D. iridis*, therefore, there is a satisfying agreement between genetic, cytological and biochemical evidence, all pointing to the conclusion that plasmodial formation begins with cell and nuclear fusion of pairs of amoebae. It would seem reasonable to suggest that the same process occurs in other species, even if a mating type system is not present. The evidence, however, does not support this idea.

4.3 Homothallic species

A life cycle is said to be 'homothallic' if cell and nuclear fusion occur between genetically identical cells. A number of Myxomycete species (10–20) have been reported to be 'homothallic' because of plasmodial formation in a high proportion of single-spore cultures. However, such observations alone certainly do not demonstrate homothallism. In the first place, it is possible that amoebae of different mating type are present in each spore and that they mate after germination. It is necessary to use single-amoeba cultures to exclude this possibility, and this has been done in only a few species.

In the few species in which plasmodial formation has been regularly found in single-amoeba cultures, however, we still cannot conclude 'homothallism'. In Fig. 4–4, five different processes are shown which could lead to the development of multinucleate cells from a colony of identical uninucleate ones. The first three of these would give rise to haploid plasmodia and the other two (including homothallism) would give diploid plasmodia. Thus measurements of nuclear DNA content or chromosome counts of amoebae and plasmodia would help us to decide which process is involved. To make further distinctions we should need to detect whether cell or nuclear fusion occur.

The formation of plasmodia in single-amoeba cultures seems to have been studied extensively and by a variety of methods in only two species, *D. nigripes* and *P. polycephalum*, and neither of these shows evidence of homothallism. In fact the evidence that there is so far indicates a different type of development. Thus there are not yet any conclusive reports of homothallism among Myxomycetes.

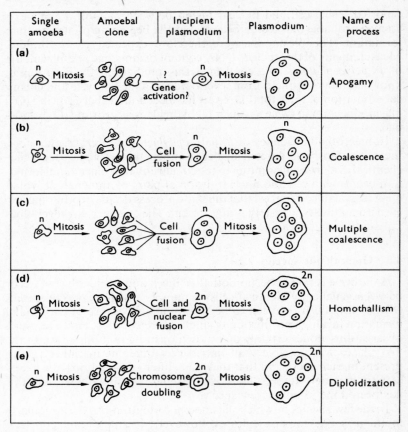

Fig. 4–4   Five processes by which a multinucleate plasmodium might develop from a clone of uninucleate amoebae (explanation in text). The ploidy of the nuclei is indicated by n (= haploid) or 2n (= diploid). Both haploid and diploid nuclei have a single large central nucleolus as indicated (see Fig. 2–1e).

4.4   Plasmodium formation without change of ploidy

In a recent study on two strains of *D. nigripes* which form plasmodia in single-amoeba cultures, approximate chromosome counts were made on samples of amoebal and plasmodial nuclei, and the results suggested that the majority of amoebae and plasmodia were of the same ploidy. Cytological observations were made on living cells of the same strains and development of multinucleate plasmodia from uninucleate cells by nuclear division *without* cell fusion was observed. Although pairs of amoebae which appeared to be fusing were often seen in the same cultures, these were never observed to give rise to plasmodia. One particularly interesting feature of this study is that the uninucleate cells which developed into plasmodia showed a number of plasmodial characteristics, in particular 'intranuclear' mitosis (Chapter 2). If the nucleus does indeed contain the same chromosome complement as an amoebal nucleus, this 'uninucleate plasmodium' must have developed from a single amoeba without cell fusion, nuclear fusion or nuclear doubling, by a mechanism which so far remains a mystery (Fig. 4–4a).

In the Colonia isolate of *P. polycephalum*, formation of plasmodia in single-amoeba cultures has occurred repeatably through many successive generations. As with *D. nigripes*, when amoebae are plated out on agar medium, colonies of amoebae are first formed but plasmodia soon develop in every one (Fig. 4–1b). The amoebae may also be cultured without plasmodial formation by frequent transfers to fresh medium. The DNA content of large numbers of amoebal and plasmodial nuclei has been very carefully measured and it is found to be the same and approximately equal to the content of heterothallic amoebae. In this study, a microdensitometer was used to measure the density of nuclei stained by the Feulgen method, which is specific for DNA. Thus it seems that haploid plasmodia are being formed in the Colonia strain but it is not known whether this is due to apogamy or to coalescence of amoebae (Fig. 4–4).

Genetic studies with the Colonia strain of *P. polycephalum* have shown that the ability to form haploid plasmodia is due to a mutant allele of the mating-type locus, which has been called 'mt_h'. All amoebae of the Colonia isolate carry this allele and are therefore capable of forming plasmodia in clones. When they are mixed with heterothallic amoebae, however, (for example a mt_1 clone) they are found to mate with them, producing heterozygous, diploid plasmodia. The amoebal progeny of these plasmodia consist of equal numbers of mt_h and heterothallic (mt_1) amoebae, as expected if mt_h and mt_1 are alleles of the same gene. Thus Colonia (mt_h) amoebae can form plasmodia either with or without nuclear fusion.

4.5 The amoebal-plasmodial-transition

When we compare amoebae and plasmodia, we see that they are different in many ways. Plasmodia are multinucleate and nuclear division occurs without cell division. In amoebae, these processes are closely linked so that a uninucleate state is maintained. In plasmodia the nuclear membrane does not break down during mitosis but in amoebae it does. Plasmodia show many morphological and physiological features which we do not see in amoebae, such as pigment, plasmodial streaming and sporulation. Plasmodia can grow on a very simple defined medium, synthesising most of the substances that they need, but amoebae apparently cannot. Plasmodia fuse readily with one another if they are of similar genotype but amoebae normally do not. From genetic studies we know that the genes controlling fusion among plasmodia do not affect mating of amoebae and that the 'mating-type' gene operating in amoebae does not affect plasmodial fusion.

It has often been assumed that plasmodial formation is initiated by cell and nuclear fusion of the amoebae to form a diploid zygote which will then rapidly grow to assume all the properties we associate with plasmodia. The studies reported above, however, indicate that plasmodium formation is not necessarily dependent on cell fusion, nuclear fusion nor even on a change of ploidy. Thus it seems that we should look elsewhere for an understanding of the developmental changes which occur.

The problem that we have to tackle is perhaps illustrated most clearly by the Colonia strain of *P. polycephalum*. When a Colonia amoeba is isolated, it multiplies to form a colony of uninucleate cells which we believe to have identical genetic content. If this colony is undisturbed, a point is reached in the life of the culture at which the majority of the amoebae become involved in forming plasmodia which soon display all the characteristic features we have listed above. Yet the DNA content and presumably the genetic content of the nuclei are still the same. The 'amoebal-plasmodial-transition' which has occurred is analogous with cell differentiation in higher organisms, where it also seems that cells with identical genetic content develop different characteristics. Such processes clearly indicate differential expression of the genes in different cell lines, but the level at which expression is controlled seems to differ in the various systems studied. In some cell lines in higher organisms there is evidence that not all the genes are transcribed; in others, there is evidence that the messenger RNA from some genes is not translated. Very little is known about the mechanisms and organization of these processes and this is an exciting area of study for many biologists at present.

In the Colonia strain, it is hoped that some understanding of the processes of differentiation in the amoebal-plasmodial-transition may come from the study of mutants. By mutagenesis of Colonia amoebae, some clones have already been isolated which fail to develop into plasmodia and these have been shown to carry mutations in genes which are not linked to the mating-type gene ('*apt*' mutants). Presumably the transition to the plasmodial state involves the 'switching-on' of genes controlling the plasmodial functions, and mutations in these genes may be expected to lead to defects or complete failure of these functions so that plasmodia do not appear. Careful investigation of mutants should give some indication of the number of functions involved and their sequential order and inter-relationships. For example, if the change from 'open' to 'closed' mitosis is the first observable stage in plasmodium formation and is controlled by different genes from those which determine that nuclear division should occur without cell division, we might expect to find some mutants which grow as a population of uninucleate cells showing closed mitosis but which fail to develop into multinucleate plasmodia. Whether we decided that we ought to call such cells 'amoebae' or 'uninucleate plasmodia' really would not matter. The important point is that we would be learning something about the genes controlling nuclear division and thus about the control of one aspect of differentiation in this organism.

4.6   The future of genetic analysis in *P. polycephalum*

Many research workers are now hoping to isolate and analyse mutants affecting processes of general biological interest such as protoplasmic streaming and the cell cycle by using the Colonia strain of *P. polycephalum*. It has been found possible to isolate mutant plasmodia in this strain by treating the amoebae with a mutagen and plating them out so that each one gives a separate clone. A mutation induced in an amoeba can be expressed in the plasmodium derived from the clone because the plasmodial nuclei are all identical and haploid. Genetic analysis can then be done by crosses with heterothallic strains. Nutritional and temperature-sensitive mutants have been isolated and analysed in this way and it is hoped that mutants affecting the cell cycle will soon be available.

Another approach which is being used in the Colonia strain is to isolate mutants in the amoebae and then to study their expression in plasmodia. The results are rather interesting in suggesting that few mutations are expressed in both stages, indicating that different sets of genes may be active in amoebae and plasmodia. The most exciting developments in this field must probably wait until *P. polycephalum* amoebae can be cultured on a defined medium so that precise biochemical comparisons of the two stages become possible. Methods for synchronizing cultures of amoebae, particularly during the amoebal-plasmodial-transition would also be extremely valuable, allowing biochemical studies of amoebae during growth and development comparable to those already done on plasmodia. In general, the most significant gains will be made when genetical and biochemical analyses can be closely correlated and progress towards this aim is slow but promising.

5 General Biology of the Cellular Slime Moulds

5.1 Life cycle of *Dictyostelium discoideum*

This species was first isolated by K. B. Raper in 1935 from forest soil in North Carolina, USA. It is not, by any means, the most common species of cellular slime mould but since it is the one most studied in research laboratories it will be described in detail. The life cycle of *D. discoideum*, like all the cellular slime moulds, is divided into two, mutually exclusive phases (Fig. 5–1). In the first, vegetative, phase the organism is found as solitary amoeboid cells (sometimes called myxamoebae to distinguish them from true amoebae) in moist, slightly acid, soils. The amoebae feed on the bacteria present in the soil and when one amoeba has eaten about 1000 bacteria it divides by mitosis into two daughter amoebae which proceed to eat more bacteria and so on. The feeding amoebae repel one another and so perfectly circular holes appear in bacterial lawns (see Appendix) as the feeding front of amoebae advances. However, when the amoebae run out of (or are deprived of) food their behaviour changes dramatically; the cells become mutually attractive and gather together to form tissue-like aggregates of cells, some of which then proceed to differentiate into new cell types which are resistant to many environmental conditions which would be lethal to the amoebae. This second, differentiation, phase of the life cycle, is thus best regarded as a device for ensuring the dispersal of the amoebae from an area where there is no food to areas where food might be plentiful.

The cellular slime moulds have two basic kinds of differentiation phase and they adopt one or the other depending on their genetic constitution and the environmental conditions. The kind of differentiation which has been most studied occurs if the non-feeding amoebae are placed on a moist, but solid, surface such as 2% agar or a Millipore filter resting on a filter paper pad saturated in buffer. Under these conditions the amoebae aggregate forming long aggregation streams consisting of tens of thousands of individual cells moving concertedly towards a central point. As the cells reach the centre of the aggregate they form a thin, finger-like mass of cells (the grex) which is usually two to four millimetres long.

When all the cells of the aggregation streams have been incorporated into the grex it bends over onto its side and begins to migrate. The

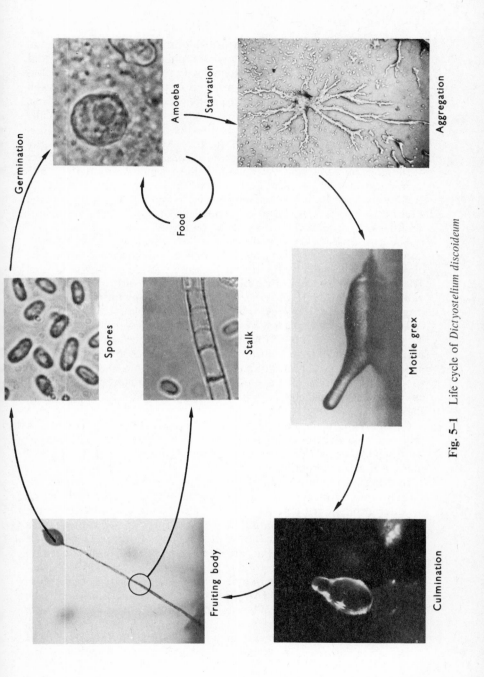

Fig. 5-1 Life cycle of *Dictyostelium discoideum*

migrating grex looks like a miniature slug and has, at the front end, a clearly defined tip. The grex as a whole moves towards the light and up temperature and humidity gradients with the tip leading. This behaviour would lead the grex to migrate out of the interstices of the soil (where aggregation occurs in Nature) and move towards the soil surface. The cells of the grex appear to move through a sheath or tube of slime which remains stationary with respect to the surface and which is thus left behind to mark the passage of the grex. After a period of time the grex stops moving, rounds up and those cells which were at the front of the grex burrow their way through those which were at the back, forming a stalk whilst those cells which were at the back move (and are lifted) up the stalk and become spores. When all the pre-spore cells have become spores, fruiting body construction is finished and the differentiation phase of the life cycle is completed.

The fruiting body consists of a spore mass which is yellow and contains about 2/3 of the total cells, and a stalk, which consists of a tube of cellulose held rigid by the turgor pressure exerted by the stalk cells. The stalk cells look very much like plant cells each with a large vacuole and thick cellulose cell walls (Fig. 5–2). The spore, in complete contrast (Fig. 5–2), has a very dense cytoplasm with no vacuole and a thick, rough cell wall. There are also a few basal disc cells (about 5–10% of the total) which remain at the foot of the stalk and which appear to be relatively undifferentiated and similar in structure to the amoebae except that they lack food vacuoles. The stalk cells become necrotic and die but the spores remain viable for years and will germinate if placed in a suitable environment when each spore gives rise to one amoeba thus completing the life cycle.

Comparison of the fine structure of the cells (Fig. 5–2) shows that there are considerable differences between the three types. This organism thus shows in a particularly clear cut fashion the general phenomenon of cell differentiation.

In addition to these processes of differentiation which lead to the formation of a fruiting body, it has also become clear that there exists an alternative developmental fate for non-growing amoebae. If the amoebae are suspended in liquid they cannot form aggregating streams as they can on a solid surface; instead they form loose spherical aggregates of cells. If the amoebae are genetically identical then nothing further happens to these loose aggregates except that the cells will eventually die since they have no food but if two, genetically distinct, populations of amoebae are mixed together in liquid suspension then large macrocysts may be formed. Whether or not two strains of D. discoideum will form a macrocyst depends on their 'mating type' and mating types may be assigned to the amoebae such that macrocysts are only formed when amoebae of different mating types are mixed together. Some

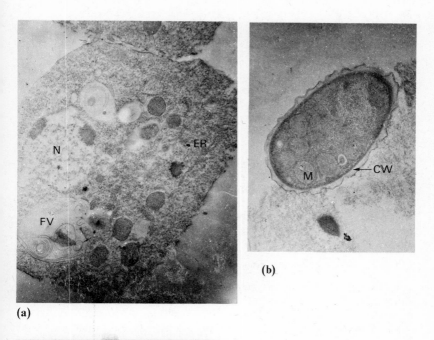

(a)

(b)

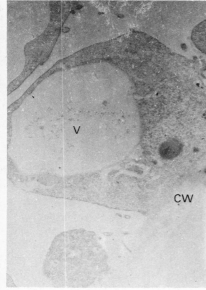

(c)

Fig. 5–2 Electron micrographs of thin sections through **(a)** an amoeba (×9000), note bacteria in food vacuoles arrowed; **(b)** a spore (×9000), and **(c)** a stalk cell (×8000). N, nucleus; ER, endoplasmic reticulum; FV, food vacuole; M, mitochondrion; V, vacuole; CW, cell wall.

species of *Dictyostelia*, however, like *D. mucoroides*, will form macro-cysts by the fusion of amoebae of the same genetic composition. Little is known of the biochemical events that occur during the formation of the macrocyst stage. Cytological studies suggest that a giant diploid cell arises as a result of fusion of two of the amoebae in the centre of the loose aggregate of cells and this grows in volume by engulfing and digesting all the other amoebae. As this engulfment proceeds, a thick multilayered cell wall is deposited around the cell mass and when engulfment has been completed the nucleus of the giant cell divides by meiosis to give rise to the multinucleated macrocyst. On germination each macrocyst gives rise to several haploid amoebae. Unfortunately it has not proved possible to find conditions which will allow germina-tion of macrocysts formed from *D. discoideum* strains with any degree of reproducibility and thus this description of the nuclear events has not yet been confirmed by the appropriate genetic analyses.

5.2 Taxonomy and ecology of the cellular slime moulds

The life cycle of *D. discoideum* was described in detail in the previous section but this species is rather rare in Nature, and has never been reported in Europe. The other species of cellular slime mould have life cycles which can be regarded, for convenience, as elaborations of that of *D. discoideum*.

5.2.1 Taxonomy of the cellular slime moulds

The Dictyostelia comprise two families, the Acytosteliidae and the Dictyosteliidae. There is only one known genus and species of the Acy-tosteliidae, *Acytostelium leptosomum*. It is the only member of the Dic-tyosteliidae which can eat yeasts as well as bacteria and its life cycle is similar to that of *D. discoideum* except that several grexes usually arise from one aggregation centre and the fruiting body consists of a spore mass of spherical spores borne at the tip of a long narrow cellulose stalk which does not contain any cells.

The family Dictyosteliidae comprises three genera, *Dictyostelium*, *Polysphondylium* and *Coenonia*. The last genus is represented by a single described species, *C. denticulata*, which was isolated by van Tieghem from decaying beans in France and has never been seen since.

The genus *Dictyostelium* is the largest of the three genera and is the one usually found in forest soils.

D. mucoroides   This species is the only one which has ever been isolated from English soils and it was the one discovered by Brefeld in Germany in 1869. Its life cycle differs from that of *D. discoideum* in that there is no clear separation of the migration stage from fruiting body construction. The developing aggregate begins to form a stalk and then as the cell mass migrates it leaves behind it a stalk instead of a collapsed slime sheath (Fig. 5–3d).

The tip shows a marked orientation towards light and will eventually turn upwards when the remaining cells become spores similar in shape to those of *D. discoideum*. Unlike *D. discoideum* there is no basal disc at the bottom of the fruiting body. Some strains of *D. mucoroides* will form macrocysts when incubated under water without the necessity of having two different mating types present, i.e. they are homothallic.

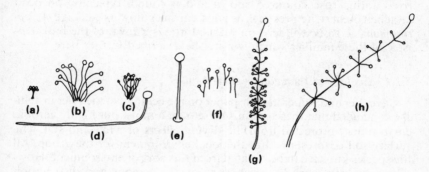

Fig. 5–3 Semidiagrammatic view of the fruiting bodies of the commoner slime moulds drawn approximately to the same scale (× 10). **(a)** *D. polycephalum*, **(b)** *D. lacteum*, **(c)** *A. leptosomum*, **(d)** *D. mucoroides* and *D. purpureum*, **(e)** *D. discoideum*, **(f)** *D. minutum*, **(g)** *P. pallidum*, **(h)** *P. violaceum*. (Redrawn from CAVENDER, J. C. and RAPER, K. B. (1965). *Amer. J. Bot.*, **52**, 302.)

Other strains will form microcysts, which consist of a cell derived from a single myxamoeba surrounded by a thin protective (probably cellulose) wall, under these circumstances. Curiously no strain has been described which can form both macrocysts and microcysts.

D. polycephalum   This species is unusual in several respects. When the amoebae aggregate they do not form streams (Fig. 5–1) so much as sheets of cells which converge about the aggregation centre and this then gives rise to a large number of separate migrating grexes. The grexes are very long and thin and show none of the usual tropisms to light and heat. When the grex ceases to move the cells pile up into a mass from which numerous stalks arise which are cemented together for most of their length. The resulting fruiting body consists of several small *D. discoideum*-like fruiting bodies (without the basal discs) bunched together (Fig. 5–3a).

The genus *Polyspondylium* comprises two known species which are the most attractive of the cellular slime moulds. They both have long thin stalks terminated by large spore masses but, in addition, have regular whorls of three or more *D. discoideum*-like fruiting bodies at intervals below the terminal spore mass. *P. pallidum* has whitish spore masses whereas the larger fruiting body of *P. violaceum* has purple spore masses (Fig. 5–3h).

The Acrasia comprise two families, the Guttulinopsidae (which contains the common *Guttulinopisis vulgaris* found on horse and cow dung) and the Acrasiae. The most common member of this family *Acrasis rosea* forms rose coloured spores and is found, typically, on dead attached plant structures such as pods, capsules, flowers etc. Like *A. leptosomum*, *A. rosea* will feed on yeasts. Little is known of the biochemistry of these families which will not be described further here.

5.2.2   Ecology of the cellular slime moulds

Cavender and Raper in their paper on the ecology of the Dictyosteliidae concluded that these organisms require a moist but highly aerobic environment protected from the drying effects of wind and sun. The surface soil of forests is thus the ideal environment for the group. All the species known can be isolated from this sort of environment. However, very little work has been done on the ecology and distribution of these organisms and since most investigators now look for them in forest soils, if there did exist a species that liked slightly alkaline, bacteria-rich soils, it might well not have been found.

The isolation of cellular slime mould species from soil samples is described in the Appendix (page 65). If this procedure is followed exactly the equivalent of $1/50 = 0.02$ g of soil is added to each isolation petri dish. Multiplication of the number of cellular slime mould clones by 50 will thus give the *absolute density* of the population in cells/g. The *frequency* (a measure of the occurrence of any species) is given by dividing the number of site occurrences of the species in question by the number of samples collected from that site (which should be at least 10) and multiplying by 100. The *relative density* of a species is calculated by dividing the number of clones of each species by the total number of clones in a population and multiplying by 100. All these calculations assume, of course, that different species have similar survival potentials under the conditions of isolation. No one has ever checked to see if this is true; can you devise an experiment which would see if it was in fact so?

J. Cavender, who, with K. B. Raper, first devised the procedure described above, surveyed the cellular slime mould populations of North America, Europe, East Africa and the Caribbean Area. Typical distribu-

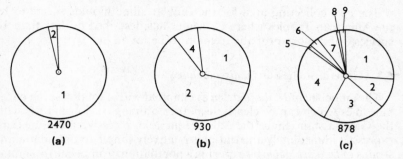

Fig. 5–4 Distribution of the cellular slime moulds in different forest soils. The area of the wedge represents the frequency with which that species occurs at each location and the number represents the total number of clones of all types obtained from one gramme of soil. (a) Oak forest near Utrecht, Holland. (b) Mixed deciduous forest near Trieste, Italy. (c) Oak forest, Pine Buff, Wisconsin, U.S.A. Key: 1, *D. mucoroides*; 2, *P. violaceum*; 3, *P. pallidum*; 4, *D. minutum*; 5, *D. polycephalum*; 6, *D. discoideum*; 7, *D. lacteum*; 8, *D. purpureum*; 9, *A. leptosomum*. (Data from CAVENDER, J. C. and RAPER, K. B. *op. cit.*, CAVENDER, J. C. (1969). *Amer. J. Bot.*, **56**, 989.)

tion patterns that he found for temperate forest soils are shown in Fig. 5–4. He found that the absolute density increased in the Spring and Autumn and that the frequency of occurrence of different species, in North America at least, varies considerably at different sites. This indicates a differential response of the various species to the microenvironment of the soil which is presumably caused, in turn, by the type of vegetation affecting the nature of the bacterial flora. However, no one has really looked in any systematic way at the relationship between the bacterial population in the leaf litter and the frequency of distribution of cellular slime mould species. It seems reasonable to imagine that different species differ in competitive ability for feeding on particular bacterial populations but, here again, no one has really looked at this question in any detail. Microorganisms have proved themselves the organisms of choice for many studies of the biochemical and molecular biological aspects of life processes and this might also prove to be true of some ecological aspects. From a theoretical point of view the predator/prey relationship between soil amoebae and soil bacteria is every bit as interesting as that between barn owls and mice, considerably easier to study and much more amenable to experimentation. In this sort of situation the factors which have made molecular biologists and biochemists turn to microorganisms (such as the ease of handling large populations of individuals, the short generation times and the wealth of detailed genetic and biochemical knowledge) apply just as well to ecologists.

The best collecting area for the cellular slime moulds seems to be the American Tropics where Cavender has described no less than 14 species—and more probably remain to be found.

5.3   Genetic studies of *D. discoideum*

The amoebae of *D. discoideum* are haploid with a total of seven chromosomes which can be clearly seen after staining cells in mitosis with the Giemsa stain. Since the cells are haploid, recessive mutations are expressed immediately and thus it is relatively simple to obtain mutant strains of *D. discoideum* by exposing a population of amoebae to a mutagenic chemical (such as N-nitro-NN-nitroso-guanidine) or a physical agent (such as ultra-violet light) and then plating out the surviving cells clonally. Careful examination reveals a variety of different types of mutant. Most obvious are those which have lost the ability to form fruiting bodies and these can be blocked at any of the intermediate stages shown in Fig. 5–1. Particularly striking are those which have lost the ability to aggregate together and thus form clear plaques in the bacterial lawn which never develop fruiting bodies. However, the most useful mutants for the purposes of genetic analysis have been those which form fruiting bodies of a different colour (white) from the normal yellow, those which are resistant to drugs, such as cycloheximide, which inhibit the growth of the wild type strains and those which are temperature sensitive. *D. discoideum* will grow and differentiate at 27 °C although the temperature optimum for both processes is 22 °C. However, after mutagenesis, W. F. Loomis at the University of California found a number of strains which would not grow at 27 °C (although they were still able to differentiate at the higher temperature).

These temperature sensitive strains have been most useful in analysing the genetic behaviour of *D. discoideum* since they provide what are known as selective markers. Thus if the amoebae of two independently isolated mutant strains (neither of which can grow at 27 °C) are mixed and allowed to form fruiting bodies, it is possible to isolate spores from these fruiting bodies which give rise to amoebae which can grow at 27 °C. However, the frequency with which such strains occur is very low (the actual figure varies from one in a thousand to one in a hundred thousand, depending on the exact conditions and the pair of mutants mixed), but they can still be detected since neither of the parental strains can grow at 27 °C and thus are completely selected against at the higher temperature. Analysis of the results of a number of mixtures of this kind suggests that during the formation of the fruiting body a para-sexual process of genetic recombination is occurring (Fig. 5–5). The first step is the fusion of two amoebae to give a binucleated cell and this seems to be quite common since many such cells can be seen in cultures of amoebae. Much more rarely these two nuclei may fuse to give rise to a diploid cell. What happens to the diploid cell depends on its genetic constitution. Some diploids seem to be very stable and give rise to numerous diploid progeny. Others, however, seem to be very unstable and lose chromosomes during division, giving rise to aneuploids (with chromosome numbers $> 7 < 16$) and eventually haploids again. The stability of the diploids seems to be a genetically controlled property of the different strains giving rise to the diploid. If the chromosomes are lost at random then, as can be seen from Fig. 5–5, the resulting haploids will either have the same genotype as the parental strains or they will have a recombinant genotype. Recombinant genotypes can also arise by a process of mitotic crossing-over which reassorts not whole chromosomes but parts of chromosomes. As the diploid cells grow and divide there is a chance that pieces of homologous chromo-

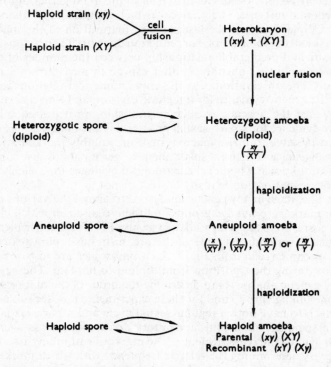

Fig. 5–5 Para-sexual cycle in *D. discoideum*. x/X and y/Y represent two un-linked genes. The genotypes of the various kinds of nuclei are in parentheses. (Redrawn from SINHA, U. K. and ASHWORTH, J. M. (1969). *Proc. Roy. Soc.*, **B173**, 531.)

somes may be interchanged. If a diploid cell in which such an interchange has occurred now gives rise to haploid progeny by chromosome loss, some of the haploid cells that result will contain a chromosome which carries characters derived initially from both of the parental strains which made up the diploid cell. The net result of mitotic crossing over events is thus the same, in the genetic sense, as occurs during meiosis but whereas chiasmata occur frequently during meiosis, cross-over events only occur, on average, during one in every 10^5 cell divisions in *D. discoideum*. This type of recombination of genetic markers is termed para-sexual because the conversion of the diploid to the haploid occurs by several stages resulting in intermediate aneuploid cells. By contrast it seems as if the formation of the macrocyst involves a genuine sexual process.

As was mentioned in section 5.1, the formation of macrocysts in *D. discoideum* only occurs as the result of mixing together amoebae of different mating types. The 'giant cell' which engulfs all the others seems to be a diploid and, at the last stage of maturation of the macrocyst there is evidence for a meiotic or reduction division. There is thus always a definite and predictable relationship between the number of the different recombinant phenotypes that can be formed during a meiotic division. The observation of such simple numerical relationships provides the genetic evidence for a meiotic division and thus a sexual process, whilst the absence of such relationships provides equally conclusive evidence for a para-sexual process. Unfortunately it has not yet proved possible to make macrocysts from suitably marked strains of *D. discoideum* and then persuade them to germinate to give amoebae, so it is not known whether the cytological evidence for meiosis during macrocyst maturation is really correct or not.

The first stage in any genetic analysis is to assign the various characters or markers to linkage groups and then to discover which chromosome corresponds to which linkage group. For these experiments it is technically very much simpler if the organism has a para-sexual cycle since, as can be seen from Fig. 5–5, chromosomes are lost or retained as units during the transition from diploid to haploid. The next stage in any genetic analysis is to determine the order of the markers within any one linkage group and for these experiments it is desirable for the organism to have both a regular sexual cycle and a para-sexual cycle, since associated with meiotic and mitotic divisions are cross-over events which lead to the reassortment of the markers within any one linkage group. By determining the relative frequency with which markers stay together or separate, it is possible to deduce how far apart they are along the chromosome. Since cross-over events occur very frequently during meiotic divisions, and very rarely during mitotic divisions, genetic markers which are very close together are best aligned by analysing the progeny of meiotic divisions whereas markers which are rather far apart are best ordered by analysing the progeny of mitotic divisions. Thus a cellular slime mould, which has both a para-sexual and a sexual method of genetic recombination, should prove to be very attractive for fundamental genetic studies, and a number of geneticists have begun work with this group of organisms.

5.4   Germination of spores and cysts and growth of amoebae

It has only recently been realized that most strains of *D. discoideum* are heterothallic and will thus form macrocysts when mixed with a strain of opposite mating type. Since these macrocysts have only recently been discovered, little is known of the conditions necessary for them to germinate and attempts to persuade macrocysts of *D. discoideum* to germinate have not been very successful to date. However, many homothallic strains of *D. mucoroides* have been described and thus the macrocysts of this species have been known for some years. In general it seems that the age of the macrocyst is the most critical factor. In some strains Nickerson and Raper reported that it was necessary to age the macrocysts for some months before they could be induced to form amoebae, although usually three or four weeks were sufficient. Light, although not essential, greatly increased the rate and frequency of macrocyst germination, especially amongst relatively young macrocysts and no requirement for nutrients of any kind could be demonstrated.

In contrast to our rather fragmentary knowledge of the conditions necessary to induce germination of the macrocysts, a lot is known about the factors which influence the germination of the spores of *D. discoideum*. Again, much of our knowledge is due to the work of Professor Raper and his colleagues at the University of Wisconsin. The spores derived from the fruiting bodies do not germinate unless they are washed well (or diluted to a density of about 10^5 spores per cm^3, which amounts to the same thing), and then given a heat shock (45 °C for 30 min) or mixed with a solution containing a mixture of amino acids (the simplest mixture that is effective contains the three compounds tryptophan, methionine and phenylalanine) at 22 °C. The need to wash the spores well is due to the fact that the spore mass contains a compound (methylguanosine) which inhibits spore germination and which is removed by extensive washing. Many fungal fruiting bodies contain such germination inhibitors and their presence means that premature germination of the spores in the fruiting body does not occur. They also ensure that the spores do not germinate until they have been washed by the rain to regions where their concentration is relatively low and thus the dispersal of the organism is made more efficient.

The amoebae seem to be quite unselective in their choice of bacterial prey and will feed on any bacterium provided that it does not secrete a thick extracellular polysaccharide capsule. It seems likely, however, that species of cellular slime mould differ in their preferences for bacteria as food and this might account for the differing proportions of the various species found in different locations (Fig. 5–4). The amoebae have a large number of special vacuoles called lysosomes which are filled with degradative enzymes and these then fuse with the food vacuoles and digest the bacteria. These food vacuoles can be clearly seen in electron micrographs of thin sections of myxamoebae (Fig. 5–2). The growth of amoebae on bacteria is described in detail in the Appendix but for many purposes it is most inconvenient to have to grow the amoebae on such a complex source of nutrients. Some years ago so-called axenic strains of *D. discoideum* were isolated which would grow on a complex mixture of nutrients in the absence of bacteria. Axenic means pertaining to the growth of a single species in the absence of living organisms or cells of any other species and the strain which was isolated at Leicester is called Ax-2.

This strain was isolated from strain NC-4, (which was in turn isolated from soil), and as a result of repeated mutations and selections it has the advantage that it will grow in a variety of media of different chemical composition. In particular, it will utilize a variety of different carbohydrates and the chemical and biochemical composition of the amoebae

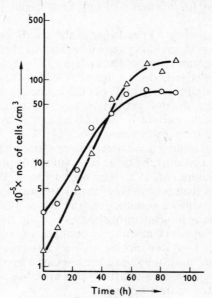

Fig. 5–6 Growth of *D. discoideum* Ax–2 in medium lacking added carbohydrate ( ) or in medium to which glucose has been added ( ). (From WATTS, D. J. and ASHWORTH, J. M. (1970). *Biochem. J.*, **119**, 171.)

alter as the medium is altered. Growth in all media is reasonably fast
with doubling times of between 8 and 12 hours (Fig. 5–6), although this
is slower than growth on *K. aerogenes* (3–4 hours doubling time). As
we shall see later, the way in which an amoeba differentiates can be
influenced by the way in which it is grown (i.e. by its initial chemical
composition) and this has also proved very useful in disentangling the
biochemical control mechanisms which regulate some aspects of differentiation. So we have a strain which can be grown in many different
ways and which will then differentiate according to several different
developmental programmes, and this allows a variety of experiments
to be done which would be impossible with other developing systems.

6 Cell Differentiation of the Cellular Slime Moulds—the Cellular Aspects

6.1 Aggregation and acrasin

The beautiful patterns formed by the aggregating amoebae on an agar surface (Fig. 6–1) attracted the attention of the early workers on the cellular slime moulds and we now probably know more about this phase of the life cycle than of any other. In fact, although patterns of the type shown in Fig. 6–1 are usually published in books (like this one) as 'typical' aggregation patterns, there are a number of other forms and patterns which may be seen, sometimes only transiently, during the aggregation process. Dr Shaffer at Cambridge University has made an extensive study of all the aggregation stages and Fig. 6–1 is taken from his review. Particularly interesting is the fact that if time lapse movie pictures are taken of the aggregation processes the cells can be seen (when the film is projected) to move in a pulsatile fashion. This is particularly obvious when the cells are plated out at a low density (about 10^5 amoebae per cm^2 of agar surface) when they form concentric wave patterns like those of Fig. 6–1 (rings) after about 7 hours' incubation at 22 °C. The cells move in a concerted and pulsatile fashion towards the centre for some hours before, usually, the precise concentric pattern breaks up and 'typical' aggregation streams (Fig. 6–1 (streams)) are formed. How can such precise, coordinated behaviour of separate, individual cells be explained? The importance of this question is that in many other developing systems complex patterns of concerted cell movements are seen and so an understanding of how amoebae aggregate might help to explain how, for example, gastrulation and neurulation occur in embryos.

Although aggregation occurs best on a solid surface it is possible to persuade amoebae to aggregate under a very thin film of water or buffer. One of the first critical experiments to be done on aggregating cells exploited this fact and by persuading amoebae to aggregate under a thin film of water which was then made to flow, it was found that the direction of the aggregating streams altered so that they pointed more 'downstream'. This experiment suggests immediately that the centre of the aggregating stream is producing a water soluble substance which is diffusing outwards and the cells in the aggregating stream are moving up concentration gradients of this substance which was given the name 'acrasin'. This interpretation was supported by the observation, made some years afterwards, that if a thin film of agar or dialysis tubing is

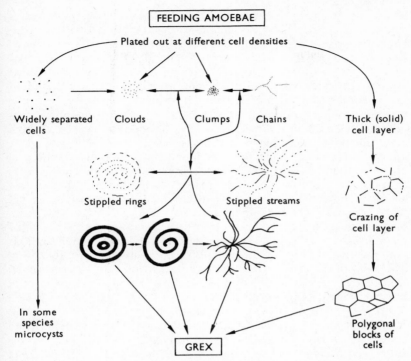

Fig. 6–1 Types of aggregation pattern seen amongst the Dictyostelliidae. Many paths of early development are reversible and these are shown by lines with arrowheads at both ends. (Redrawn after SHAFFER, B. M. (1962). *Adv. in Morph.*, **2**, 109.)

placed over an aggregating stream and amoebae placed on the surface of the agar or tubing then they will form an aggregation stream whose shape is identical in all respects to that of the lower stream. Since cells cannot pass through such films but small molecules can, the implication is that the aggregation centres and their satellite streams are secreting a substance to which the second population of amoebae are responding even though there can, in this case, be no direct cell-cell contacts. A lot of effort then went into attempts to isolate and characterize this substance 'acrasin'. This soon proved a very difficult task, and the situation was not helped by the fact that the ability of an acrasin preparation to attract suitable amoebae often 'disappeared' in ways and after treatments which suggested that the cells were also excreting an enzyme called, therefore, 'acrasinase', which destroyed 'acrasin'. Thus acrasin containing solutions could not be kept for very long without their

activity diminishing. One well known and often tried trick in these circumstances is to see if an alternative source of material with acrasin-
like activity can be found which is not contaminated with acrasinase.
A number of such sources were found, including urine, but the one that
provided the answer in the end was spent bacterial culture fluids. Bacteria can be grown very simply and enormous quantities of the liquid
in which they have been grown can be obtained cheaply and readily.
Chemical fractionation of this material in Professor Bonner's laboratory at Princeton gave a material which, although it had a very high
acrasin activity, was obviously still grossly contaminated with other
materials. However, it had some properties which resembled those of
nucleotides and this led the Princeton group to see which nucleotides
amongst those known to occur in living cells had any acrasin-like
activity. One of the compounds that they tried was a substance cyclic-
AMP (Fig. 6–2) which had been isolated some years previously from
mammalian cells, where it is known to be involved, as the so-called
second messenger, in a variety of hormone mediated responses. This
material proved to be identical in all respects with the natural acrasin.
The story of the discovery of the chemical identity of acrasin, which
extends over nearly 30 years and has involved work by biologists of
many nationalities, is typical of many such investigations. Progress was,
at first, very slow as the complexities of the problem were discovered
but patient work over many years eventually produced a reliable assay
system which led to increasingly pure preparations and finally an in-
spired and brilliant guess by a young student at Princeton solved the
problem. Problems, however, are rarely solved as neatly in scientific
investigations as they are in detective stories and there is an ironic and
unexpected postscript to this work. When Bonner and his group came
to compare the properties of the material isolated from spent bacterial
culture fluids with cyclic-AMP, some discrepancies appeared. The more
pure the material from the bacterial culture fluids became, the less like

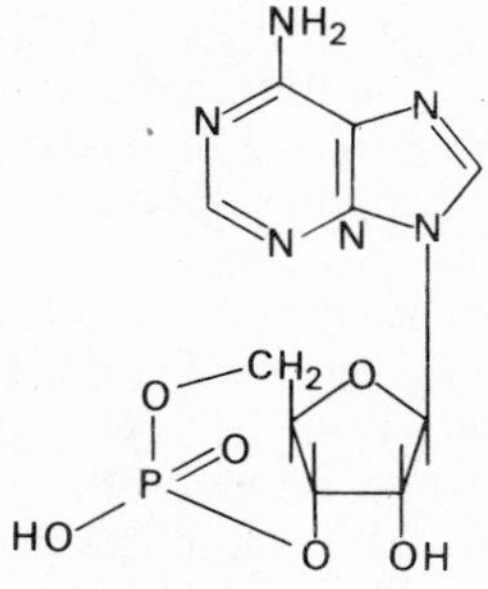

Fig. 6–2   Formula of cyclic-AMP.

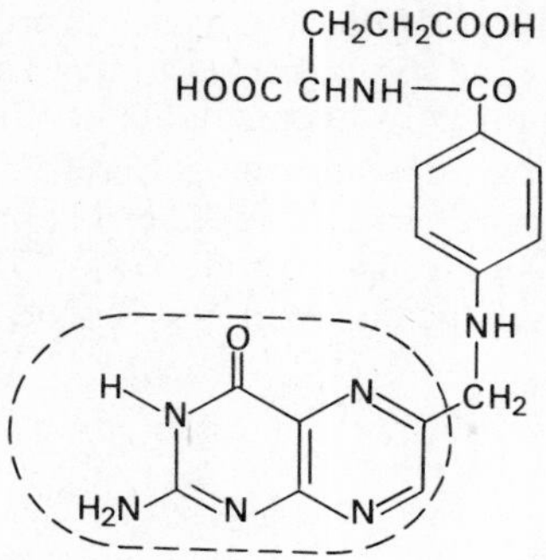

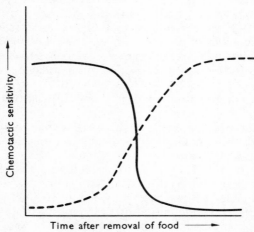

Fig. 6–3 Formula of folic acid. The portion ringed (pterin group) seems to be more important in attracting amoebae than the rest of the molecule.

cyclic-AMP it appeared until eventually it had to be accepted that there are two types of compound, not one, which attract amoebae. The first of these, acrasin, is made and excreted by the amoebae themselves and is identical with cyclic-AMP but the second, which is made and excreted by bacteria, appears to be either folic acid (Fig. 6–3) or compounds closely related to it chemically. Amoebae which have just stopped feeding on bacteria are extremely sensitive to folic acid but not very sensitive to cyclic-AMP. The longer the amoebae are prevented from feeding the more sensitive they become to cyclic-AMP and the less sensitive they are to attraction by folic acid (Fig. 6–4).

Fig. 6–4 Change in sensitivity of amoebae to folic acid (——) and cyclic-AMP (---).

Although the discovery of the chemical nature of acrasin provides a useful object lesson in the way chemical analysis and biological insight interact, it does not really carry us much further, of itself, in an attempt to account for the aggregation patterns of Fig. 6–1. However, being able to obtain acrasin in a bottle from commercial suppliers led to the rapid identification of acrasinase as a cyclic-AMP phosphodiesterase, an enzyme which catalyses the conversion of cyclic-AMP to AMP:

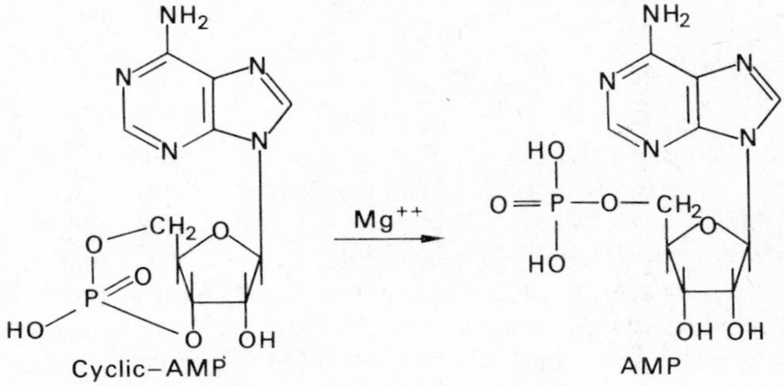

Fig. 6–5 Reaction catalysed by acrasinase.

and we now know that amoebae make not only acrasinase but they also make and secrete another protein which acts as a very specific inhibitor of the acrasinase. The amount of cyclic-AMP that there is around an amoebae will depend on the rate at which cyclic-AMP is being synthesized by that, and other, amoebae and the rate at which the cyclic-AMP is being destroyed by the acrasinase. It is clear that the amoebae can control both these rates. These further facts still do not account, however, for the movement of amoebae towards aggregation centres. At the moment the best description of aggregation would be that:

(1) the aggregation centre periodically emits a pulse of cyclic-AMP; and

(2) nearby amoebae receive a cyclic-AMP signal and respond by moving towards the signalling centre and, by emitting their own pulse, relay the cyclic-AMP pulse.

Direct evidence for (1) has come from the finding that amoebae respond to periodic injections of cyclic-AMP from a micropipette just as they do to an aggregation centre provided the period is appropriate and (2) is necessary for the formation of the aggregation streams.

(3) After receiving and responding to the cyclic-AMP signal the amoebae must become insensitive to another cyclic-AMP signal for a period of time known as the refractory period; and

(4) during the refractory period the signalling pulse of cyclic-AMP must be removed by the acrasinase.

These two properties of the amoebae are needed for the cells to 'sense' the direction in which the next signal is coming from and for them to respond to it in a unidirectional way. A remarkable combination of mathematical analysis and careful analysis of movie films of aggregation has shown that by choosing appropriate values for the time taken to relay a cyclic-AMP signal, for the refractory period and knowing the rate of movement of the individual cells the existence and appearance of the whorls and concentric circle patterns of aggregation can be accounted for and, indeed, predicted.

As the cells enter the aggregating streams they change their appearance markedly and seem to have a definite front end and a definite back end. This is because, during the aggregation phase, the cell membrane changes in such a way that adhesive sites are formed, probably in patches, and the cells link in a head to tail fashion using these adhesive or binding sites. Thus at the end of the aggregation phase a truly multicellular entity, the grex, has been formed in which the component cells adhere strongly one to another and this goes on to form the fruiting body.

6.2   Pattern formation

During cell differentiation there occurs, in general, the structural and functional specialization of individual cells from one of a number of common basic cell types, each of which is competent to develop in several different ways. For example, the mesenchyme cells of the embryonic limb bud may become, amongst other things, either muscle or cartilage cells. The processes whereby those biochemical and cytological features which distinguish muscle cells from cartilage cells are acquired by the differentiating cell, are largely intracellular and recent advances in molecular biology have done much to elucidate what the processes might be and how they might be controlled. However, what is so remarkable about the whole process is not that muscles and cartilage appear but that they appear at the right time, at the right place and in the right relative proportions. Pattern formation is the name given to the process which ensures that the correct spatial and proportional relationships are established and maintained between differentiating cells. The nature of this process is one of the most perplexing in developmental biology and we appear to have in the cellular slime moulds a tantalizingly simple and clear cut example. The end result of the developmental cycle is a fruiting body in which the spore and stalk cells are in such a spatial relationship to one another that a fruiting body of a definite shape is formed (Fig. 5–1) and, further, we know as a result of detailed and careful analysis that the number of spore cells bears a constant and proportional relationship to the number of stalk cells despite wide variations in the overall size of the fruiting body. The fruiting body thus appears to be the result of a process of pattern formation just as precise and perhaps of the same form as that which gives rise to a vertebrate limb.

Although manifestation of the cellular pattern as the appearance of stalk and spore cells only occurs during fruiting body formation, there is a lot of evidence which suggests that the pattern arises much earlier in the developmental cycle. The first evidence of this came from some elegant grafting experiments done by Professor Raper. He had shown that if amoebae were grown on the bacterium *Serratia marcesans* they accumulated the red pigment characteristic of this bacterium in their vacuoles. When they aggregated they therefore produced, not translucent, colourless, grexes but red ones. If a normal, colourless, motile grex is now placed side by side with a red one and the front 1/3 of the grexes are exchanged a composite grex is obtained. Since the grex just consists of a mass of cells encased in a slime sheath, these operations can easily be done with a pair of fine needles, watchmaker's forceps or even a loop of (human) hair. So long as the agar on which the grexes are placed is reasonably moist, the pieces can easily be pushed about using needles and healing of the grafts and reconstitution of the grexes takes place readily. When such composite grexes form fruiting bodies, it is found that the front 1/3 of the cells always form the stalk and the back 2/3 the spores. Thus there is at least a functional distinction between the front and the back cells of a grex. Detailed examination of the grex cells has also revealed that the back cells have a characteristic structure, the pre-spore vacuole, absent from the front cells. A number of other differences in, for example, density, enzyme composition and adhesiveness to a substratum have been reported between these two groups of cell and it is quite clear that the pattern which will be expressed at fruiting body construction is already present in the grex. However, it is equally clear that this pattern is not irreversibly impressed on the grex cells for, if a grex is cut into thirds and the pieces allowed to develop in isolation, then each piece can form a fruiting body containing the correct spore : stalk cell ratio (Fig. 6–6). The piece which contains the tip of the grex will only form a fruiting body with the correct proportions if it is allowed to migrate; if it forms a fruiting body immediately after surgery, then a structure which is mostly, if not entirely, stalk is produced. It thus seems as if the tip is in some way involved in organizing the rest of the grex and that movement is necessary in order for the correct proportions to be established if a grex finds itself, as a result of surgery, with a tip developed for an assemblage of cells which was initially much larger. A number of suggestions have been put forward

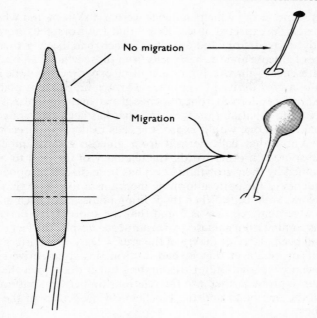

Fig. 6–6   Result of cutting a motile grex of *D. discoideum* into thirds.

to account for these facts but as yet no one coherent explanation can be accepted. Since to carry out the grafting experiments requires little except a dissecting microscope and standard school biology apparatus, the field would appear to be wide open for an ingenious young researcher who is not bemused by all the 'facts' known to experts. Probably only two such facts are going to be significant in the long run. The first of these is that Professor Bonner has shown that when amoebae are placed at very low cell densities on agar plates containing a high concentration of cyclic-AMP (10^{-3}M), so that aggregation cannot occur, the isolated cells differentiate to forms cells which look much like isolated stalk cells. The second is that we know that the eventual fate of a cell can be affected by the way in which the amoebae from which it was derived were grown. In these experiments amoebae were grown in different conditions, such as in media with and without glucose (Fig. 5–6), and then mixed together before being placed on Millipore filters to form fruiting bodies. It was found that those amoebae which had been grown in the glucose containing medium formed spores in preference to those which had been grown in the absence of glucose. This implies that the two types of amoebae had 'sorted out' one from another in the grex and this could be shown to be the case by cutting up grexes and seeing which amoebae were in the front and which were at the back. As expected, it was found that the front cells were predominantly derived from the amoebae which had been grown in the absence of glucose and the back cells from those which had been grown in the presence of glucose. If now a grafting experiment is done in which the tip of a grex derived from glucose-grown amoebae is placed on a decapitated grex derived from amoebae grown in the absence of glucose, an interesting situation results. From what has been said above, in this composite grex one would expect the cells in the tip to become stalk but we know that cells derived from glucose grown amoebae will become spores. In other words, the tendency of the cells to 'sort out' (determined by their growth history) has been placed in opposition to the tendency of the pattern forming mechanism to ensure that tip cells will become stalk cells. When the fruiting bodies from such composite grexes were analysed, it was found that the spores were derived from the cells coming from glucose grown amoebae despite the fact that these cells had been placed at the tip of the grex. Clearly the pattern forming mechanism, whatever that is, had 'lost' in this competitive situation. We have no explanation for this finding but it does open the way for attempts to probe further into the relationship between pattern formation on the one hand and the biochemical properties of the cells on the other.

Although it is clear that the pattern forming mechanism in the grex orders cells into front cells and back cells, it is not so clear that this pattern forming process is necessarily the same as that which orders the spore and stalk cells in the fruiting body. When amoebae are gently shaken in buffer they will, in 18 hours or so, form large balls of cells which can be 0.5 mm in diameter. These cells cannot undergo differentiation into spores and stalks because they are immersed in liquid but if they are placed on an agar surface they will 'sprout' fruiting bodies which look, after 6–8 hours, rather like a bunch of miniature daffodils. Careful examination shows that numerous protuberances appear on the top of the initially spherical surface of the cell mass and these look very much like the tip of a grex. A fruiting body is then formed from each protuberance without the intermediate formation of a proper grex, motile or otherwise. This suggested to Professor Gerisch that in some way a liquid/air interface was necessary for tip formation and that the tip organized the pattern of the fruiting body with or without the necessary intervention of the motile grex stage. To test this point he placed such spherical balls of cells, not now on an agar surface, but in the interstices of a nylon net (e.g. a nylon stocking) that had been soaked in water. Now the balls of cells were exposed to *two* liquid/air interfaces and sure enough they produced fruiting bodies after incubation in a humid atmosphere at 22 °C directed away from both sides of the nylon

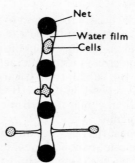

Fig. 6–7 Differentiation of an aggregate of *D. discoideum* cells exposed to two air/water interfaces. (From GERISCH, G. (1960). *Arch. EntwMech. Org.* **152**, 632.)

support (Fig. 6–7). This demonstrates the importance of liquid/air interfaces and the importance of the tip without, however, giving us much clue as to how the pattern formation occurs.

It is clear that although the problem of pattern formation in the cellular slime mould appears simple, the basic mechanism is at the moment complex enough to baffle us and it is probably true that until we can really understand the slime mould we are unlikely to be able to understand how similar pattern forming mechanisms work during the embryological development of higher organisms. However, these are not by any means the only unknown processes that occur in developing embryos. Of key importance, and connected with pattern formation, is the problem of morphogenesis.

6.3   Morphogenesis

Morphogenesis is the name given to the largely mechanical process by which the form of the organism and of its tissues is generated. Morphogenesis thus involves the coordinated movement of cells, sometimes in large groups (e.g. sheets) and sometimes as individuals. The most dramatic example of this in most higher animals is gastrulation when the mesoderm and endoderm move through the blastopore into the inside of what was the blastula whilst the ectoderm spreads to cover them. Clearly these movements and rearrangements must be closely coordinated both in space and time and a basic problem of morphogenesis is how such movements occur and how they are controlled and coordinated.

According to the above definition, the aggregation stage of the cellular slime mould is the first one in which morphogenesis occurs. Since the aggregating cells, at any rate during the early stages of aggregation, form only a unicellular layer on the substratum, the morphogenetic movements are particularly easy to see and to analyse. In the preceding section we discussed one important aspect of the aggregation process, chemotaxis, but of equal importance from the point of view of the final shape of the aggregation stream is the interactions the cells make with one another and with the substratum. Clearly the final shape of the aggregation stream will depend on the relative strengths of these inter-actions and the manner in which the cell movement responds to the acrasin (cyclic-AMP) signal.

During the feeding phase the amoebae do not adhere strongly to each other and that slight amount of cohesion which does occur is inhibited by concentrations of the chelating agent EDTA (ethylene-diamino-tetra-acetic acid) low enough to leave the cells themselves unharmed. During the aggregation phase, however, the mutual adhesiveness of the cells increases markedly due to changes in the chemical composition of the outer surface of the cell membrane. New proteinaceous adhesive sites appear and these make the cells stick together even in the presence of low concentrations of EDTA. These sites do not appear to be distri-buted at random over the cell surface but are so arranged that the cells, when they enter aggregation streams (Fig. 5–1), align themselves 'head-to-tail'. We would very much like to know the exact composition of these novel adhesive sites, how they are synthesized, how they can stick cells to one another and yet allow the cells so stuck to move and so on. These sorts of problems are also evident in other developmental situations and answers will probably provide the key to a number of baffling problems besides that posed by aggregation in *D. discoideum*.

In the migrating grex the cells are piled on top of one another several layers deep. The cells are thus not so easy to see, or follow, and a number of new morphogenetic problems arise such as how the slug-like shape is generated and maintained and how the slug moves as an integrated whole. The grex is surrounded by a slime sheath which remains station-ary relative to the substratum as the cell mass moves through it. This sheath collapses as the last cells move forward and is left behind as a slimy record of the path taken by the grex. It is clear, therefore, that the slime sheath must be continually synthesized at the tip and the tip and slime sheath must play a critical role in polarizing grex movement. The importance of the tip can be shown most clearly by grafting tips into various parts of the grex. When a number of tips are grafted later-ally each is followed by the cells immediately behind it so that several, smaller, grexes are produced (Fig. 6–8). However, the overall direction of movement of a grex cannot be reversed by grafting a tip onto the back of another grex.

A baffling problem is also presented by the attempts to understand how the cells at the top of the grex move at the same speed as those on the bottom, as they do. If those on top use the backs of those beneath

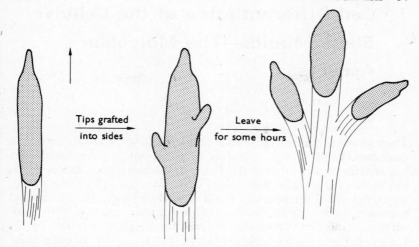

Fig. 6–8   The effect of grafting tips laterally into a migrating grex. Movement is in the arrowed direction. (Redrawn from RAPER, K. B. (1940). *J. Elisha Mitchell Scient. Soc.*, **56**, 241.)

as a substratum they will move forward relative to the true substratum at twice the speed of those in the lower layer. If the cell surfaces remain stationary whilst the cell moves forward by continuously synthesizing a new surface at the front and resolving the old one at the back this difficulty is removed but direct experimental tests of this hypothesis have given conflicting results. These problems, and others, have been discussed by D. R. Garrod who has given references to the original work. Many of these experiments can be quite easily performed by students with the aid of a binocular microscope and simple glass needles and, since so little is known about the mechanisms of these processes, this is one field where an ingenious student can still contribute to knowledge without needing enormous amounts of expensive equipment.

7 Cell Differentiation of the Cellular Slime Moulds—The Molecular Aspects

Three kinds of temperature sensitive mutant of *D. discoideum* have been described (Table 1), corresponding to mutations in genes specifically required for growth processes (GTS), or developmental processes (DTS) or in genes required for both stages of the life cycle (TS). Temperature sensitivity arises because, as a result of gene mutation, the amino-acid sequence of a protein has been altered in a manner altering its thermostability. Thus the implication of these results (first obtained by Dr Loomis of the University of California) is that during the cell differentiation processes new kinds of proteins are specifically required to catalyse the synthesis of the novel structures which characterize the stalk and spore cells. The fruiting body contains many polysaccharides not found in the amoebae and direct evidence that novel proteins are needed for their synthesis has come from the study of changes in specific activity of three enzymes concerned in carbohydrate metabolism (Fig. 7–1). These patterns of change are affected in various developmental mutants in a way which suggests that the alterations in specific activity are coupled to the morphological changes. At the moment much effort is being directed at understanding the mechanisms underlying the changes shown in Fig. 7–1 and, although several different theories have been put forward, we really have little idea of the control mechanisms involved. Two interesting observations have been made, however, which suggest that these mechanisms are neither simple nor straightforward. We noticed that if amoebae were grown axenically in the presence of glucose (so that they had a high cellular glycogen content) and then allowed to differentiate, they produced more trehalose than if they were

Table 1 Temperature sensitive mutant strains of *D. discoideum*.

Strain*	Growth at 27°C	Development at 27°C
Wild type	+	+
GTS	−	+
DTS	+	−
TS	−	−

* all strains grow and develop normally at 22°C

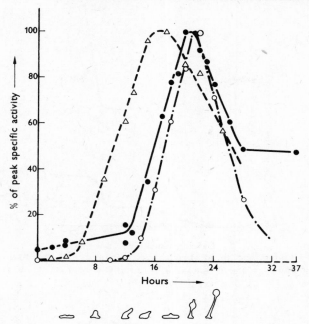

Fig. 7–1 Changes in the specific activity of three enzymes concerned in carbohydrate metabolism during the differentiation of *D. discoideum*. Trehalose-6-phosphate synthase (△), UDP-glucose pyrophosphorylase (●) and UDP-galactose: polysaccharide transferase (○). (From ROTH, R., ASHWORTH, J. M. and SUSSMAN, M. (1968). *Proc. natn. Acad. Sci. U.S.A.*, **59**, 1235.)

grown in the absence of glucose (and thus had a low cellular glycogen content initially), although there was no difference in the time taken by the two cell populations to differentiate or in the morphological sequence they followed. The differences in trehalose content between the two cell populations were considerable (Fig. 7–2) and yet, rather to our surprise, there was absolutely no difference in the activity per cell of the enzyme (Fig. 7–3), which is known to catalyse the rate limiting stages in trehalose synthesis—trehalose synthase.

This experiment shows that there is no necessary connection between the amount of enzyme which a differentiating cell manufactures and the extent to which that enzyme will be used.

The second experiment was done by Drs Newell and Sussman and consisted of examining the effects of cell-cell contacts on the synthesis of trehalose synthase and the other enzymes of Fig. 7–1. When developing cell assemblies are suspended in dilute buffer and subjected to gentle

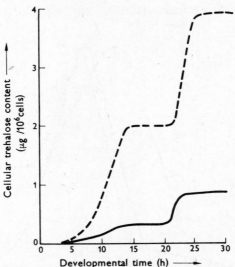

Fig. 7–2 Trehalose content of cells derived from amoebae containing a high (− −) or low (− −) glycogen content initially. (From HAMES, B. D. and ASHWORTH, J. M. (1974). *Biochem. J.*, **142**, 317.)

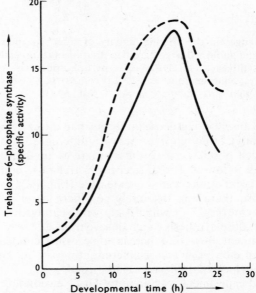

Fig. 7–3 Trehalose synthase activities in the cells whose trehalose contents are shown in **Fig. 7–2**. (From HAMES, B. D. and ASHWORTH, J. M., (1974). *Biochem. J.*, **142**, 317.)

sheer forces (such as those produced by sucking a cell suspension up and down a pipette), the cell-cell contacts are broken and an essentially homogeneous suspension of single cells results. If this suspension is now redeposited on a Millipore filter the cells recapitulate their previous developmental history in half an hour or so and form cell assemblies which are minute replicas of the morphological stage that they had reached prior to the dissociation. These minute assemblies then proceed to form minute fruiting bodies in much the same time that they would have formed normal sized fruiting bodies had they been left undisturbed. Determinations of the effects of dissociation on the specific activity of trehalose synthase are shown in Fig. 7–4. It can be seen that disruption of the cell-cell interactions leads to a reinitiation of the synthesis of this enzyme and the final appearance of approximately double the amount of enzymic activity formed in the undissociated controls. The amount of trehalose actually formed by fruiting bodies from the dissociated samples is, if anything, lower than that formed by the controls and so, once again, there is no correspondence between the amount

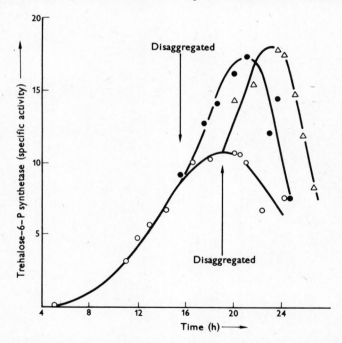

Fig. 7–4 Changes in trehalose synthase activity in cells dissociated at 15 and 18 h and then redeposited on filters (●, △). Undissociated controls (○) (From NEWELL, P. C., FRANKE, J. and SUSSMAN, M. (1972), *J. Mol. Biol.*, **63**, 373.)

of an enzyme synthesized and the 'use' that is made of that activity. Thus the 'rules' for regulating the amount of any enzyme formed during development do not seem to involve maximizing the efficiency of some of that enzyme but rather involve cell-cell and other interactions about which we know very little at the moment. *D. discoideum* does, however, once again pose the question of 'what regulates the synthesis of developmentally important proteins?' in a peculiarly straightforward fashion and at the current rate of progress we should soon know much more of these controls.

Appendix

Note (i): Sterile technique should be observed throughout.

Note (ii): All agar concentrations given are for Japanese agar and may need adjustment if a different kind is used. A fairly hard surface is required, especially for amoebae. Agar plates should not be stored for many days after pouring as they become too dry.

A.1 Culture methods for *Physarum polycephalum*

A.1.1 Plasmodia

MEDIA (a) *Oat agar:* 8 g rolled oats, 1 g agar, 100 cm^3 water. Autoclave dry oats and water-agar separately (15 p.s.i./15 min), mix while agar is still hot and pour.

(b) *McArdle medium:* Add to water in this order: 3.6 g citric acid 2H$_2$O, 0.054 g FeCl$_2$.H$_2$O, 0.54 g MgSO$_4$.7H$_2$O, 0.54 g CaCl$_2$.2H$_2$O, 0.076 g MnCl$_2$.4H$_2$O, 0.03 g ZnSO$_4$.7H$_2$O, 9 g tryptone, 9 g dextrose (anhydrous), 1.4 g yeast extract, 1.8 g K$_2$HPO$_4$, water to 1000 cm^3. Adjust pH to 4.6 with 30% KOH. Autoclave 15 p.s.i./15 min. Make separate solution of hematin (Koch-Light Laboratories, Colnbrook, Bucks.): 0.05 g in 100 cm^3 1% NaOH, autoclave, store at 4°C. Add 1 cm^3 hematin to 100 cm^3 medium just before use. To prepare plates, mix medium with equal quantity 2.25% melted agar and pour.

(c) *Minimal defined medium:* (DM-1). See DEE, J., WHEALS, A. E. and HOLT, C. E. (1973). *Genet. Res.*, Cambridge, **21,** 87–101.

TO CULTURE PLASMODIA Transfer a block ~1 cm^2 to fresh medium with spatula. Incubate 25°–28 °C. Transfer about once per week. Most plasmodia will age and die after several months in continuous culture; so keep stocks as sclerotia, spores or amoebae.

MIGRATION AGAR To free plasmodia of contaminants, allow migration over 1.5% agar, pH 4.6 for 1–2 days.

SCLEROTIA Plasmodia allowed to remain on the surface of medium in the dark, after growth is complete, often form sclerotia which have a dry and papery texture. These may be peeled off and stored dry in screw cap bottles. Germinate sclerotia by placing on fresh medium for about one week. However, sclerotia produced on oat agar may have low viability.

A.1.2   Sporulation

Place plasmodial culture in daylight or white fluorescent light after incubation in dark for about one week, a day or two after the nutrient supply appears to be exhausted. Avoid direct sunlight. If difficulty is experienced with sporulation on oat agar, try reducing oats to a few flakes/plate.

A.1.3   Culture from spores

Place spores in $1–2\,cm^3$ distilled water, crush clumps with glass rod, leave overnight or several hours at room temperature. Resuspend, count and plate as for amoebae (*A.1.4*).

A.1.4.   Amoebae

MEDIUM   *Liver infusion agar (LIA):* 1 g liver infusion powder (Oxoid), 15 g agar, 1 l distilled water.

BACTERIAL SUSPENSION   Keep stock cultures of bacteria (e.g. *Escherichia coli*) on any suitable nutrient agar slopes (e.g. Oxoid) and subculture regularly. To make suspension for feeding amoebae, wash the bacterial growth from a 24-hour culture into distilled water and shake. (Approx. $2\,cm^2$ of surface culture in $2\,cm^3$ water will make a suitable 'milky' suspension.)

PLATING OUT AMOEBAE FOR SINGLE COLONIES   Make suspension of amoebae in water; mix with pasteur pipette or Vortex mixer; count with haemacytometer; dilute to give approximately 500 amoebae/cm^3. (If no haemacytometer is available, use a 3 mm diameter loopful of water to transfer amoebae from a lawn to $2\,cm^3$ water and mix thoroughly. This will give approximately 500 cells/cm^3). Transfer $0.1\,cm^3$ of this suspension to each LIA plate (9 cm Petri dish), with $0.05\,cm^3$ bacterial suspension, and spread evenly with bent glass rod. Incubate $25°–26\,°C$. Colonies ('plaques') become visible after 4 days. *Note:* clear colonies will be seen only if the surface of the agar dries within a few hours of spreading.

CULTURING AMOEBAE   Dip small wire loop in water; touch on amoebal colony without breaking droplet; draw across surface of fresh LIA slope (or plate) inoculated with $0.05\,cm^3$ bacterial suspension (surface can be wet when incubating). Incubate $25°–26\,°C$. Subculture to fresh slopes at intervals of a few weeks.

A.1.5   Plasmodial formation

MEDIA   (a) *Dilute McArdle (DMA):* Add 7 cm³ McArdle medium (including hematin) to 100 cm³ melted 1.5% agar. Pour into plates.
(b) *LIA + PABA(LIAP):* Add p-amino-benzoic acid to LIA to give final concentration of 57μg/cm³.

CROSSING HETEROTHALLIC AMOEBAE   Inoculate DMA or LIAP plate with 0.05 cm³ bacterial suspension as 'puddle' in centre. Place one loopful of amoebae from each heterothallic strain in wet puddle and mix. Incubate 25°–26 °C. Plasmodia appear after 3–7 days. Transfer small plasmodia to McArdle medium to which streptomycin has been added (250 μg/cm³) to kill *E. coli.*

MATING-TYPE TESTS   Use plates poured previous day so that surface is dry. Inoculate each plate with several spots of bacterial suspension by means of a loop; allow to soak in; inoculate each spot with different combination of amoebal strains. Inspect daily after 3 days' incubation. Cut out spots when plasmodia form to avoid migration.

PLASMODIA FROM COLONIA AMOEBAE   Inoculate DMA or LIAP plates with Colonia amoebae and bacteria as for crosses above. To obtain plasmodia from single colonies, plate out 20–50 Colonia amoebae per plate on DMA or LIAP as in (*A.1.4*).

A.1.6   Isolation and culture of other Myxomycetes

Some of the media and methods used for *P. polycephalum* may be suitable for Myxomycetes isolated from nature. For further information on collection and culture of other species, see Chapter 11 of GRAY and ALEXOPOULOS, and the article by CARLILE in *Methods in Microbiology* (see Further Reading).

A.2   Methods for handling the cellular slime moulds

A.2.1   Isolation of cellular slime moulds from soil

Select damp areas in woodlands and take a mixture of the leaf litter and the surface soil layer. About 10 g of damp soil is mixed with 90 cm³ of sterile water and shaken vigorously to break up any lumps and ensure an even suspension of solid matter. 5 cm³ of the resulting suspension is then added to 7.5 cm³ of sterile water and 0.5 cm³ of this is then mixed with about 0.4 cm³ of an overnight culture of *K. aerogenes* (see Section A.2.2b). This mixture is now spread evenly on petri dishes containing 'isolation medium' (see Section *A.2.2a*). The point of this medium is that it is not rich enough to support much growth of the fungal spores inevitably present in the soil sample but it should provide sufficient nutrients to form a fairly thick bacterial lawn in which the slime moulds can grow. After 3–4 days clones should appear in the bacterial lawn and since each species grows at a different rate, it is essential to keep checking the plates for clones for a week or more.

A.2.2   Growth of the cellular slime moulds

(a) *Isolation medium* contains (in grams per litre) glucose 1; bacteriological peptone 1; K_2HPO_4 1; KH_2PO_4 1.5 and $MgSO_4$ 1; agar 20.
(b) *SM medium* contains (in grams per litre) glucose 10; bacteriological peptone 10; yeast extract 1; $MgSO_4$ 1; KH_2PO_4 1.5; K_2HPO_4 1 and agar 20. An inoculum of the bacterial food *Klebsiella aerogenes* (formerly called *Aerobacter aerogenes*) is prepared by adding a few cells of the bacterium to $5\,cm^3$ of SM medium lacking the agar contained in a sterile test tube. Overnight incubation of this tube will give a thick bacterial suspension and a few drops of this are mixed with a sterile suspension of spores and the mixture spread over the surface of a petri dish containing about $30\,cm^3$ of SM medium so that the surface is completely covered. After 48 h incubation at 22° it should be possible to see translucent regions in the bacterial lawn and after 3 to 4 days fruiting bodies should be visible. If large amounts of amoebae are required it is best to add an inoculum of about 10^5 spores to each plate and to harvest before the bacterial lawn has been completely cleared.

A.2.3   Harvesting of amoebae

When translucent areas are visible in the bacterial lawn and before all the bacteria have been eaten, $10\,cm^3$ of cold water are pipetted into the petri dish. The lawn of amoebae and bacteria can then be scraped off the agar surface with a glass rod (or the fingers if you are not too squeamish and remember to wash well afterwards) and the agar further washed with $2 \times 10\,cm^3$ portions of cold water. The suspension of amoebae and bacteria is then centrifuged at $500\,g$ for 5 minutes, the pellet resuspended in $30\,cm^3$ of cold water, recentrifuged and the pellet further washed with $2 \times 30\,cm^3$ portions of cold water and finally resuspended in cold water at a cell density of 10^8 cells/cm³.

A.2.4   Studies of differentiating amoebae

(a) The washed suspension of amoebae is spread on freshly prepared petri dishes containing 2% agar in 50 mM phosphate buffer pH 6.5. At low cell densities beautiful aggregation streams can be seen but the development is not very synchronous.

(b) For highly synchronous development amoebae are best deposited on filter paper discs which themselves rest on filter paper pads moistened with sufficient buffer (1.5 g/l KCl, 0.5 g/l $MgCl_2.6H_2O$ and 0.5 g/l streptomycin sulphate in 50 mM phosphate buffer, pH 6.5) to maintain a visible drop of liquid near the pads.

A.3   Sources of supply of the slime moulds

The Curator, The Culture Collection, Commonwealth Mycological Institute, Ferry Lane, Kew, Surrey will supply sclerotia of a heterothallic plasmodium of *P. polycephalum* (mating type 1 × mating type 2). Amoebae derived from spores will segregate 1 : 1 for mating types 1 and 2. Spores of *D. discoideum* can also be obtained from the same source. A charge of about £2.00 is made for each culture.

The American Type Culture Collection, (12301 Parklawn Drive, Rockland, Maryland 20852, USA) also holds stocks of slime moulds and charges $25 for each culture issued.

Further Reading

Myxomycetes

1 GRAY, W. D. and ALEXOPOULOS, C. J. (1968). *Biology of the Myxomycetes*. Ronald Press, New York.
2 HUTTERMAN, A. (Ed.) (1973). *Physarum polycephalum—Object pf Research in Cell Biology*. Gustav Fischer Verlag, Stuttgart. This volume, containing 8 reviews on aspects of *Physarum* research, is also published as an issue of the journal *Berichte der Deutschen Botanischen Gesellschaft*, (**86**, Parts 1–4, 1973).
3 CARLILE, M. J. (1971). Chap. IX, Myxomycetes and other slime moulds. In *Methods in Microbiology*, Vol. 4, Booth, C. Academic Press, London and New York.
4 CARLILE, M. J. (1970). Nutrition and chemotaxis in the Myxomycete *Physarum polycephalum J. gen. Microbiol.*, **63**, 221–6.
5 DEE, J. (1975) Slime moulds in biological research. *Science Progress, Oxford*, **62** (in press).
6 ING, B. (1968). *A census catalogue of British Myxomycetes*. The Foray Committee of the British Mycological Society.

Cellular slime moulds

1 BONNER, J. T. (1967). *The Cellular Slime Moulds*. Princeton University Press, Princeton. This book is a very readable and comprehensive account of all the work done with the cellular slime moulds before 1966.
2 LOOMIS, W. F. (1975). *Dictyostelium discoideum: A Developmental System*. Academic Press, New York. This book is a comprehensive account of the work done in the period 1966–74.